ALGEBRA I FOR BEGINNERS

The Ultimate Step by Step Guide to Acing Algebra I

By

Reza Nazari

Copyright © 2023
Effortless Math Education

All rights reserved. No part of this publication may be reproduced, stored in a retrieval system, or transmitted in any form or by any means, electronic, mechanical, photocopying, recording, scanning, or otherwise, except as permitted under Section 107 or 108 of the 1976 United States Copyright Ac, without permission of the author.

All inquiries should be addressed to:
info@effortlessMath.com
www.EffortlessMath.com

ISBN: 978-1-63719-264-1

Published by: **Effortless Math Education**

for Online Math Practice Visit www.EffortlessMath.com

Welcome to
Algebra I Prep
2023

Thank you for choosing Effortless Math for your Algebra I preparation and congratulations on making the decision to take the Algebra I course! It's a remarkable move you are taking, one that shouldn't be diminished in any capacity.

That's why you need to use every tool possible to ensure you succeed on the test with the highest possible score, and this extensive study guide is one such tool.

Algebra I for Beginners is designed to be comprehensive and cover all the topics that are typically covered in an Algebra I course. It provides clear explanations and examples of the concepts and includes practice problems and quizzes to test your understanding of the material. The textbook also provides step-by-step solutions to the problems, so you can check your work and understand how to solve similar problems on your own.

Additionally, this textbook is written in a user-friendly way, making it easy to follow and understand even if you have struggled with math in the past. It also includes a variety of visual aids such as diagrams, graphs, and charts to help you better understand the concepts.

Algebra I for Beginners is flexible and can be used to supplement a traditional classroom setting, or as a standalone resource for self-study. With the help of this comprehensive textbook, you will have the necessary foundation to master the material and succeed in the Algebra I course.

Effortless Math's Algebra I Online Center

Effortless Math Online Algebra I Center offers a complete study program, including the following:

- ✓ Step-by-step instructions on how to prepare for the Algebra I test
- ✓ Numerous Algebra I worksheets to help you measure your math skills
- ✓ Complete list of Algebra I formulas
- ✓ Video lessons for all Algebra I topics
- ✓ Full-length Algebra I practice tests
- ✓ And much more…

No Registration Required.

Visit **EffortlessMath.com/Algebra1** to find your online Algebra I resources.

How to Use This Book Effectively

Look no further when you need a study guide to improve your math skills to succeed on the Algebra I test. Each chapter of this comprehensive guide to the Algebra I will provide you with the knowledge, tools, and understanding needed for every topic covered on the course.

It's imperative that you understand each topic before moving onto another one, as that's the way to guarantee your success. Each chapter provides you with examples and a step-by-step guide of every concept to better understand the content that will be on the course. To get the best possible results from this book:

- **Begin studying long before your test date**. This provides you ample time to learn the different math concepts. The earlier you begin studying for the test, the sharper your skills will be. Do not procrastinate! Provide yourself with plenty of time to learn the concepts and feel comfortable that you understand them when your test date arrives.
- **Practice consistently**. Study Algebra I concepts at least 20 to 30 minutes a day. Remember, slow and steady wins the race, which can be applied to preparing for the Algebra I test. Instead of cramming to tackle everything at once, be patient and learn the math topics in short bursts.
- Whenever you get a math problem wrong, **mark it off, and review it later** to make sure you understand the concept.
- Start each session by **looking over the previous material.**
- Once you've reviewed the book's lessons, **take a practice test at the back of the book** to gauge your level of readiness. Then, review your results. Read detailed answers and solutions for each question you missed.
- **Take another practice test** to get an idea of how ready you are to take the actual exam. Taking the practice tests will give you the confidence you need on test day. Simulate the Algebra I testing environment by sitting in a quiet room free from distraction. Make sure to clock yourself with a timer.

Looking for more?

Visit EffortlessMath.com/Algebra1 to find hundreds of Algebra I worksheets, video tutorials, practice tests, Algebra I formulas, and much more.

Or scan this QR code.

No Registration Required.

Contents

CHAPTER

1 Fundamental and Building Blocks

Math topics that you'll learn in this chapter:

- ☑ Adding and Subtracting Integers
- ☑ Multiplying and Dividing Integers
- ☑ Translate a Phrase into an Algebraic Statement
- ☑ Order of Operations
- ☑ Integers and Absolute Value
- ☑ Proportional Ratios
- ☑ Similarity and Ratios
- ☑ Percent Problems
- ☑ Percent of Increase and Decrease
- ☑ Discount, Tax, and Tip
- ☑ Simple Interest
- ☑ The Distributive Property
- ☑ Approximating Irrational Numbers

Adding and Subtracting Integers

- Integers include zero, counting numbers, and the negative of the counting numbers. $\{\ldots, -3, -2, -1, 0, 1, 2, 3, \ldots\}$
- Add a positive integer by moving to the right on the number line. (You will get a bigger number)
- Add a negative integer by moving to the left on the number line. (You will get a smaller number)
- Subtract an integer by adding its opposite.

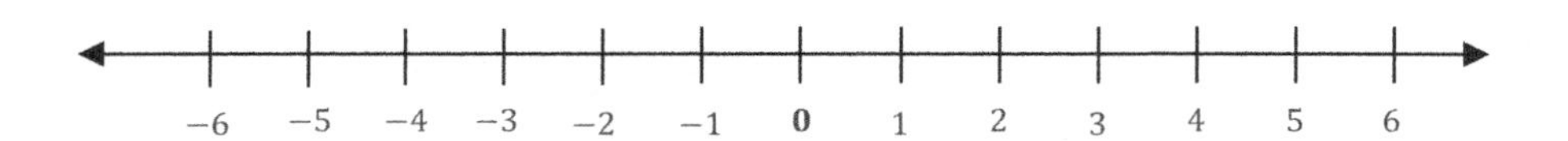

Examples:

Example 1. Solve. $(-2) - (-8) =$

Solution: Keep the first number and convert the sign of the second number to its opposite. Change subtraction into addition. Then:

$(-2) + 8 = 6.$

Example 2. Solve. $4 + (5 - 10) =$

Solution: First, subtract the numbers in brackets, $5 - 10 = -5$.
Then: $4 + (-5) = \rightarrow$ change addition into subtraction:

$4 - 5 = -1.$

Example 3. Solve. $(9 - 14) + 15 =$

Solution: First, subtract the numbers in brackets, $9 - 14 = -5$.
Then:

$-5 + 15 = \rightarrow -5 + 15 = 10.$

Example 4. Solve. $12 + (-3 - 10) =$

Solution: First, subtract the numbers in brackets, $-3 - 10 = -13$.
Then: $12 + (-13) = \rightarrow$ change addition into subtraction:
$12 - 13 = -1.$

Multiplying and Dividing Integers

Use the following rules for multiplying and dividing integers:

- $(negative) \times (negative) = positive$
- $(negative) \div (negative) = positive$
- $(negative) \times (positive) = negative$
- $(negative) \div (positive) = negative$
- $(positive) \times (positive) = positive$
- $(positive) \div (negative) = negative$

Examples:

Example 1. Solve. $3 \times (-4) =$

Solution: Use this rule: $(positive) \times (negative) = negative.$
Then:

$(3) \times (-4) = -12.$

Example 2. Solve. $(-3) + (-24 \div 3) =$

Solution: First, divide -24 by 3, the numbers in brackets, use this rule: $(negative) \div (positive) = negative$. Then: $-24 \div 3 = -8$.

$(-3) + (-24 \div 3) = (-3) + (-8) = -3 - 8 = -11.$

Example 3. Solve. $(12 - 15) \times (-2) =$

Solution: First, subtract the numbers in brackets,

$12 - 15 = -3 \rightarrow (-3) \times (-2) =$

Now use this rule: $(negative) \times (negative) = positive \rightarrow (-3) \times (-2) = 6.$

Example 4. Solve. $(12 - 8) \div (-4) =$

Solution: First, subtract the numbers in brackets,

$12 - 8 = 4 \rightarrow (4) \div (-4) =$

Now use this rule: $(positive) \div (negative) = negative.$

$\rightarrow (4) \div (-4) = -1.$

Translate a Phrase into an Algebraic Statement

Translating keywords and phrases into algebraic expressions:

- Addition: plus, more than, the sum of, etc.
- Subtraction: minus, less than, decreased, etc.
- Multiplication: times, product, multiplied, etc.
- Division: quotient, divided, ratio, etc.

Examples:

Translate each phrase into an algebraic statement.
Example 1. 14 times the sum of 7 and x.

The sum of 7 and x: $7 + x$. Times means multiplication.
Then: $14 \times (7 + x) = 14(7 + x)$.

Example 2. Six more than a number is 16.

"More than" means plus. "a number" is x.
Then: $6 + x = 16$.

Example 3. 17 times the sum of 4 and x.

The sum of 4 and x: $4 + x$. Times means multiplication. Then: $17 \times (4 + x) = 17(4 + x)$.

Example 4. Five more than a number is 24.

More than mean plus a number $+x$.
Then: $5 + x = 24$.

Example 5. 15 times the sum of 2 and x.

The sum of 2 and x: $2 + x$. Times means multiplication. Then: $15 \times (2 + x)$.

Example 6. Three more than a number is 11.

"More than" means plus. "a number" is x.
Then: $3 + x = 11$.

Order of Operations

- In Mathematics, "operations" are addition, subtraction, multiplication, division, exponentiation (written as b^n), and grouping.
- When there is more than one math operation in an expression, use PEMDAS: (To memorize this rule, remember the phrase "Please Excuse My Dear Aunt Sally".)
  - ✓ Parentheses
  - ✓ Exponents
  - ✓ Multiplication and Division (from left to right)
  - ✓ Addition and Subtraction (from left to right)

Examples:

Example 1. Calculate. $(2 + 6) \div (2^2 \div 4) =$

Solution: First, simplify inside parentheses:

$$(8) \div (4 \div 4) = (8) \div (1).$$

Then: $(8) \div (1) = 8$.

Example 2. Solve. $(6 \times 5) - (14 - 5) =$

Solution: First, calculate within parentheses:

$$(6 \times 5) - (14 - 5) = (30) - (9).$$

Then: $(30) - (9) = 21$.

Example 3. Calculate. $-4[(3 \times 6) \div (9 \times 2)] =$

Solution: First, calculate within parentheses:

$$-4[(18) \div (9 \times 2)] = -4[(18) \div (18)] = -4[1].$$

Multiply -4 and 1. Then: $-4[1] = -4$.

Example 4. Solve. $(28 \div 7) + (-19 + 3) =$

Solution: First, calculate within parentheses:

$$(28 \div 7) + (-19 + 3) = (4) + (-16).$$

Then: $(4) - (16) = -12$.

Integers and Absolute Value

- The absolute value of a number is its distance from zero, in either direction, on the number line. For example, the distance of 9 and -9 from zero on the number line is 9.
- The absolute value of an integer is the numerical value without its sign. (Negative or positive)
- The vertical bar is used for absolute value as in $|x|$.
- The absolute value of a number is never negative; it only shows, "how far the number is from zero".

Examples:

Example 1. Calculate. $|14-2| \times 5 =$

Solution: First, solve $|14-2|, \rightarrow |14-2| = |12|$, the absolute value of 12 is 12, $|12| = 12$.

Then: $12 \times 5 = 60$.

Example 2. Solve. $\frac{|-24|}{4} \times |5-7| =$

Solution: First, find $|-24| \rightarrow$ the absolute value of -24 is 24. Then: $|-24| = 24$, $\frac{24}{4} \times |5-7| =$.

Now, calculate $|5-7|, \rightarrow |5-7| = |-2|$, the absolute value of -2 is 2: $|-2| = 2$. Then: $\frac{24}{4} \times 2 = 6 \times 2 = 12$.

Example 3. Solve. $|8-2| \times \frac{|-4\times 7|}{2} =$

Solution: First, calculate $|8-2|, \rightarrow |8-2| = |6|$, the absolute value of 6 is 6, $|6| = 6$. Then: $6 \times \frac{|-4\times 7|}{2}$.

Now calculate $|-4 \times 7|, \rightarrow |-4 \times 7| = |-28|$, the absolute value of -28 is 28, $|-28| = 28$. Then: $6 \times \frac{28}{2} = 6 \times 14 = 84$.

Proportional Ratios

- Two ratios are proportional if they represent the same relationship.
- A proportion means that two ratios are equal. It can be written in two ways: $\frac{a}{b} = \frac{c}{d}$ $\quad a : b = c : d$.
- The proportion $\frac{a}{b} = \frac{c}{d}$ can be written as $a \times d = c \times b$.

Examples:

Example 1. Solve this proportion for x. $\frac{2}{5} = \frac{6}{x}$

Solution: Use cross multiplication: $\frac{2}{5} = \frac{6}{x} \rightarrow 2 \times x = 6 \times 5 \rightarrow 2x = 30$.

Divide both sides by 2 to find x:

$$x = \frac{30}{2} \rightarrow x = 15.$$

Example 2. If a box contains red and blue balls in ratio of $3:5$ red to blue, how many red balls are there if 45 blue balls are in the box?

Solution: Write a proportion and solve. $\frac{3}{5} = \frac{x}{45}$

Use cross multiplication: $3 \times 45 = 5 \times x \rightarrow 135 = 5x$.

Divide to find x:

$$x = \frac{135}{5} \rightarrow x = 27.$$

There are 27 red balls in the box.

Example 3. Solve this proportion for x. $\frac{4}{9} = \frac{16}{x}$

Solution: Use cross multiplication: $\frac{4}{9} = \frac{16}{x} \rightarrow 4 \times x = 9 \times 16 \rightarrow 4x = 144$.

Divide both sides by 4 to find x:

$$x = \frac{144}{4} \rightarrow x = 36.$$

Example 4. Solve this proportion for x. $\frac{5}{7} = \frac{20}{x}$

Solution: Use cross multiplication: $\frac{5}{7} = \frac{20}{x} \rightarrow 5 \times x = 7 \times 20 \rightarrow 5x = 140$.

Divide to find x:

$$x = \frac{140}{5} \rightarrow x = 28.$$

bit.ly/37GHQxp

Find more at

Similarity and Ratios

- Two figures are similar if they have the same shape.
- Two or more figures are similar if the corresponding angles are equal, and the corresponding sides are in proportion.

Examples:

Example 1. The following triangles are similar. What is the value of the unknown side?

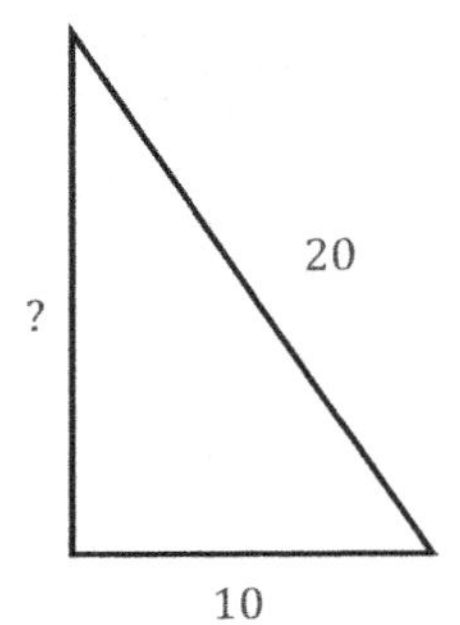

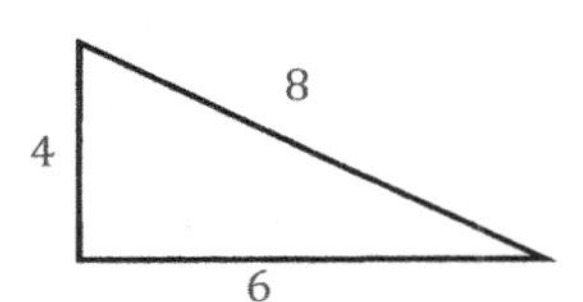

Solution: Find the corresponding sides and write a proportion.

$\frac{8}{20} = \frac{6}{x}.$

Now, use the cross-product to solve for x:

$\frac{8}{20} = \frac{6}{x} \rightarrow 8 \times x = 20 \times 6 \rightarrow 8x = 120.$

Divide both sides by 8. Then:

$8x = 120 \rightarrow x = \frac{120}{8} \rightarrow x = 15.$

The missing side is 15.

Example 2. Two rectangles are similar. The first is 5 feet wide and 15 feet long. The second is 10 feet wide. What is the length of the second rectangle?

Solution: Let's put x for the length of the second rectangle. Since two rectangles are similar, their corresponding sides are in proportion. Write a proportion and solve for the missing number.

$\frac{5}{10} = \frac{15}{x} \rightarrow 5x = 10 \times 15 \rightarrow 5x = 150 \rightarrow x = \frac{150}{5} = 30.$

The length of the second rectangle is 30 feet.

bit.ly/2KKKmc
Find more at

Percent Problems

- Percent is a ratio of a number and 100. It always has the same denominator, 100. The percent symbol is "%".
- Percent means "per 100". So, 20% is $\frac{20}{100}$.
- In each percent problem, we are looking for the base, the part, or the percent.
- Use these equations to find each missing section in a percent problem:
  - $Base = Part \div Percent$
  - $Part = Percent \times Base$
  - $Percent = Part \div Base$

Examples:

Example 1. What is 20% of 40?

Solution: In this problem, we have the percent (20%) and the base (40) and we are looking for the "part".
Use this formula:

$$Part = Percent \times Base.$$

Then:

$$Part = 20\% \times 40 = \frac{20}{100} \times 40 = 0.20 \times 40 = 8.$$

The answer: 20% of 40 is 8.

Example 2. 25 is what percent of 500?

Solution: In this problem, we are looking for the percent.
Use this equation:

$$Percent = Part \div Base \rightarrow Percent = 25 \div 500 = 0.05 = 5\%.$$

Then: 25 is 5 percent of 500.

Example 3. 80 is 20 percent of what number?

Solution: In this problem, we are looking for the base. Use this equation:

$$Base = Part \div Percent \rightarrow Base = 80 \div 20\% = 80 \div 0.20 = 400.$$

Then: 80 is 20 percent of 400.

Percent of Increase and Decrease

- Percent of change (increase or decrease) is a mathematical concept that represents the degree of change over time.
- To find the percentage of increase or decrease:
  1. $New\ Number - Original\ Number$,
  2. $(The\ result \div Original\ Number) \times 100$.
- Or use this formula: $percent\ of\ change = \frac{new\ number - original\ number}{original\ number} \times 100$.
- Note: If your answer is a negative number, then this is a percentage decrease. If it is positive, then this is a percentage increase.

Examples:

Example 1. The price of a shirt increases from $\$30$ to $\$36$. What is the percentage increase?

Solution: First, find the difference: $36 - 30 = 6$.
Then:

$$(6 \div 30) \times 100 = \frac{6}{30} \times 100 = 20.$$

The percentage increase is 20%. It means that the price of the shirt increased by 20%.

Example 2. The price of a table decreased from $\$50$ to $\$35$. What is the percentage of the decrease?

Solution: Use this formula:

$$Percent\ of\ change = \frac{new\ number - original\ number}{original\ number} \times 100.$$

Then:

$$\frac{35-50}{50} \times 100 = \frac{-15}{50} \times 100 = -30.$$

The percentage decrease is 30. (The negative sign means a percentage decrease.) Therefore, the price of the table decreased by 30%.

Chapter 1: Practices

Find each sum or difference.

1) $-9+16=$
2) $-18-6=$
3) $-24+10=$
4) $30+(-5)=$
5) $15+(-3)=$
6) $(-13)+(-4)=$
7) $25+(3-10)=$
8) $12-(-6+9)=$
9) $5-(-2+7)=$
10) $(-11)+(-5+6)=$
11) $(-3)+(9-16)=$
12) $(-8)-(13+4)=$
13) $(-7+9)-39=$
14) $(-30+6)-14=$
15) $(-5+9)+(-3+7)=$
16) $(8-19)-(-4+12)=$
17) $(-9+2)-(6-7)=$
18) $(-12-5)-(-4-14)=$

Solve.

19) $3\times(-6)=$
20) $(-32)\div 4=$
21) $(-5)\times 4=$
22) $(25)\div(-5)=$
23) $(-72)\div 8=$
24) $(-2)\times(-6)\times 5=$
25) $(-2)\times 3\times(-7)=$
26) $(-1)\times(-3)\times(-5)=$
27) $(-2)\times(-3)\times(-6)=$
28) $(-12+3)\times(-5)=$
29) $(-3+4)\times(-11)=$
30) $(-9)\times(6-5)=$
31) $(-3-7)\times(-6)=$
32) $(-7+3)\times(-9+6)=$
33) $(-15)\div(-17+12)=$
34) $(-3-2)\times(-9+7)=$
35) $(-15+31)\div(-2)=$
36) $(-64)\div(-16+8)=$

Write an algebraic expression for each phrase.

37) 11 multiplied by x = ____________________
38) 18 divided by x = ____________________
39) The square of 15. ____________________
40) The difference between ninety–six and y. ____________
41) The difference between x and 32 is 15. ________________

Evaluate each expression.

42) $3 + (2 \times 5) =$
43) $(5 \times 4) - 7 =$
44) $(-9 \times 2) + 6 =$
45) $(7 \times 3) - (-5) =$
46) $(-8) + (2 \times 7) =$
47) $(9 - 6) + (3 \times 4) =$
48) $(-19 + 5) + (6 \times 2) =$
49) $(32 \div 4) + (1 - 13) =$
50) $(-36 \div 6) - (12 + 3) =$
51) $(-16 + 5) - (54 \div 9) =$
52) $(-20 + 4) - (35 \div 5) =$
53) $(42 \div 7) + (2 \times 3) =$
54) $(28 \div 4) + (2 \times 6) =$
55) $2[(3 \times 3) - (4 \times 5)] =$
56) $3[(2 \times 8) + (4 \times 3)] =$
57) $2[(9 \times 3) - (6 \times 4)] =$
58) $4[(4 \times 8) \div (4 \times 4)] =$
59) $-5[(10 \times 8) \div (5 \times 8)] =$

Find the answers.

60) $|-5| + |7 - 10| =$
61) $|-4 + 6| + |-2| =$
62) $|-9| + |1 - 9| =$
63) $|-7| - |8 - 12| =$
64) $|9 - 11| + |8 - 15| =$
65) $|-7 + 10| - |-8 + 3| =$
66) $|-12 + 6| - |3 - 9| =$
67) $5 + |2 - 6| + |3 - 4| =$
68) $-4 + |2 - 6| + |1 - 9| =$
69) $\frac{|-42|}{7} \times \frac{|-64|}{8} =$
70) $\frac{|-100|}{10} \times \frac{|-36|}{6} =$
71) $|4 \times (-2)| \times \frac{|-27|}{3} =$
72) $|-3 \times 2| \times \frac{|-40|}{8} =$
73) $\frac{|-54|}{6} - |-3 \times 7| =$
74) $\frac{|-72|}{8} + |-7 \times 5| =$
75) $\frac{|-121|}{11} + |-6 \times 4| =$
76) $\frac{|(-6)\times(-3)|}{9} \times \frac{|2\times(-20)|}{5} =$
77) $\frac{|(-3)\times(-8)|}{6} \times \frac{|9\times(-4)|}{12} =$

Solve each proportion.

78) $\frac{3}{2} = \frac{9}{x} \rightarrow x =$ ____

79) $\frac{7}{2} = \frac{x}{4} \rightarrow x =$ ____

80) $\frac{1}{3} = \frac{2}{x} \rightarrow x =$ ____

81) $\frac{1}{4} = \frac{5}{x} \rightarrow x =$ ____

82) $\frac{9}{6} = \frac{x}{2} \rightarrow x =$ ____

83) $\frac{3}{6} = \frac{5}{x} \rightarrow x =$ ____

84) $\frac{7}{x} = \frac{2}{6} \rightarrow x =$ ____

85) $\frac{2}{x} = \frac{4}{10} \rightarrow x =$ ____

86) $\frac{3}{2} = \frac{x}{8} \rightarrow x =$ ____

87) $\frac{x}{6} = \frac{5}{3} \rightarrow x =$ ____

88) $\frac{3}{9} = \frac{5}{x} \rightarrow x =$ ____

89) $\frac{4}{18} = \frac{2}{x} \rightarrow x =$ ____

90) $\frac{6}{16} = \frac{3}{x} \rightarrow x =$ ____

91) $\frac{2}{5} = \frac{x}{20} \rightarrow x =$ ____

92) $\frac{28}{8} = \frac{x}{2} \rightarrow x =$ ____

93) $\frac{3}{5} = \frac{x}{15} \rightarrow x =$ ____

94) $\frac{2}{7} = \frac{x}{14} \rightarrow x =$ ____

95) $\frac{x}{18} = \frac{3}{2} \rightarrow x =$ ____

96) $\frac{x}{24} = \frac{2}{6} \rightarrow x =$ ____

97) $\frac{5}{x} = \frac{4}{20} \rightarrow x =$ ____

98) $\frac{10}{x} = \frac{20}{80} \rightarrow x =$ ____

99) $\frac{90}{6} = \frac{x}{2} \rightarrow x =$ ____

Solve each problem.

100) Two rectangles are similar. The first is 8 *feet* wide and 32 *feet* long. The second is 12 *feet* wide. What is the length of the second rectangle? ________________

101) Two rectangles are similar. One is 4.6 *meters* by 7 *meters*. The longer side of the second rectangle is 28 *meters*. What is the other side of the second rectangle? __________________

Effortless Math Education

Solve each problem.

102) What is 15% of 60? ___

103) What is 55% of 800? ___

104) What is 22% of 120? ____

105) What is 18% of 40? ___

106) 90 is what percent of 200? ___%

107) 30 is what percent of 150? ___%

108) 14 is what percent of 250? ___%

109) 60 is what percent of 300? ___%

110) 30 is 120 percent of what number? ___

111) 120 is 20 percent of what number? ___

112) 15 is 5 percent of what number? ___

113) 22 is 20% of what number? ___

Solve each problem.

114) Bob got a raise, and his hourly wage increased from $15 to $21. What is the percent increase?

115) The price of a pair of shoes increases from $32 to $36. What is the percentage increase?

116) At a coffee shop, the price of a cup of coffee increased from $1.35 to $1.62. What is the percent increase in the cost of the coffee?

117) TRGBA $45 shirt now selling for $36 is discounted by what percent?

118) Joe scored 30 out of 35 marks in Algebra, 20 out of 30 marks in science, and 58 out of 70 marks in mathematics. In which subject his percentage of marks is best?

119) Emma purchased a computer for $420. The computer is regularly priced at $480. What was the percent discount Emma received on the computer?

120) A chemical solution contains 15% alcohol. If there is 54 *ml* of alcohol, what is the volume of the solution?

Find the selling price of each item.

121) Original price of a computer: $600. Tax: 8%, Selling price: $_______

122) Original price of a laptop: $450. Tax: 10%, Selling price: $_______

123) Nicolas hired a moving company. The company charged $500 for its services, and Nicolas gave the movers a 14% tip. How much does Nicolas's tip the movers? $_________

124) Mason has lunch at a restaurant and the cost of his meal is $40. Mason wants to leave a 20% tip. What is Mason's total bill, including tip? $_________

Determine the simple interest for the following loans.

125) $1,000 at 5% for 4 years. $____

126) $400 at 3% for 5 years. $___

127) $240 at 4% for 3 years. $___

128) $500 at 4.5% for 6 years. $___

Solve.

129) A new car, valued at $20,000, depreciates at 8% per year. What is the value of the car one year after purchase? $_____________

130) Sara put $7,000 into an investment yielding 3% annual simple interest; she left the money in for five years. How much interest does Sara get at the end of those five years? $_____________

Use the distributive property to simplify each expression.

131) $2(6 + x) =$ ________

132) $5(3 - 2x) =$ ________

133) $7(1 - 5x) =$ ________

134) $(3 - 4x)7 =$ ________

135) $6(2 - 3x) =$ ________

136) $(-1)(-9 + x) =$ ________

137) $(-6)(3x - 2) =$ ________

138) $(-x + 12)(-4) =$ ________

139) $(-2)(1 - 6x) =$ ________

140) $(-5x - 3)(-8) =$ ________

Find the approximation of each.

141) $\sqrt{44} \approx$ ____

142) $\sqrt{72} \approx$ ____

143) $\sqrt{27} \approx$ ____

144) $\sqrt{92} \approx$ ____

Find the approximation of each and locate them approximately on a number line diagram.

145) $\sqrt{7} \approx$ ____

-6 -5 -4 -3 -2 -1 0 1 2 3 4 5 6

146) $\sqrt{30} \approx$ ____

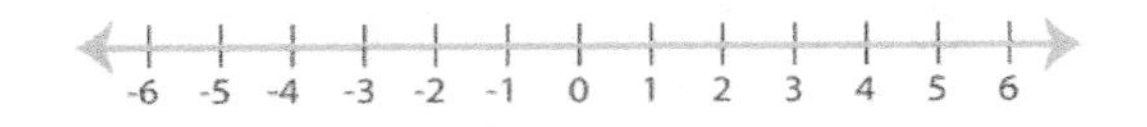

Chapter 1: Answers

1) 7
2) -24
3) -14
4) 25
5) 12
6) -17
7) 18
8) 9
9) 0
10) -10
11) -10
12) -25
13) -37
14) -38
15) 8
16) -19
17) -6
18) 1
19) -18
20) -8
21) -20
22) -5
23) -9
24) 60
25) 42
26) -15
27) -36
28) 45
29) -11
30) -9
31) 60
32) 12
33) 3
34) 10
35) -8
36) 8
37) $11x$
38) $\frac{18}{x}$
39) 15^2
40) $96 - y$
41) $x - 32 = 15$
42) 13
43) 13
44) -12
45) 26
46) 6
47) 15
48) -2
49) -4
50) -21
51) -17
52) -23
53) 12
54) 19
55) -22
56) 84
57) 6
58) 8
59) -10
60) 8
61) 4
62) 17
63) 3
64) 9
65) -2
66) 0
67) 10
68) 8
69) 48
70) 60
71) 72
72) 30
73) -12
74) 44
75) 35
76) 16
77) 12
78) 6
79) 14
80) 6
81) 20
82) 3
83) 10
84) 21
85) 5
86) 12
87) 10
88) 15
89) 9
90) 8
91) 8
92) 7
93) 9
94) 4
95) 27
96) 8
97) 25
98) 40
99) 30
100) $48\ feet$
101) $18.4\ meters$
102) 9
103) 440
104) 26.4
105) 7.2

Effortless Math Education

106) 45%
107) 20%
108) 5.6%
109) 20%
110) 25
111) 600
112) 300
113) 110
114) 40%
115) 12.5%
116) 20%
117) 20%
118) Algebra
119) 12.5%
120) $360ml$
121) $648.00
122) $495.00
123) $70.00
124) $48.00
125) $200.00
126) $60.00
127) $28.80
128) $135.00
129) $18,400
130) $1,050
131) $2x + 12$
132) $-10x + 15$
133) $-35x + 7$
134) $-28x + 21$
135) $-18x + 12$
136) $-x + 9$
137) $-18x + 12$
138) $4x - 48$
139) $12x - 2$
140) $40x + 24$
141) 6.6
142) 8.5
143) 5.2
144) 9.6

145) 2.6

-6 -5 -4 -3 -2 -1 0 1 2 3 4 5 6

146) 5.5

-6 -5 -4 -3 -2 -1 0 1 2 3 4 5 6

CHAPTER

2 Exponents and Variables

Math topics that you'll learn in this chapter:

- ☑ Multiplication Property of Exponents
- ☑ Division Property of Exponents
- ☑ Powers of Products and Quotients
- ☑ Zero and Negative Exponents
- ☑ Negative Exponents and Negative Bases
- ☑ Scientific Notation
- ☑ Addition and Subtraction in Scientific Notation
- ☑ Multiplication and division in Scientific Notation

Multiplication Property of Exponents

- Exponents are shorthand for repeated multiplication of the same number by itself. For example, instead of 2×2, we can write 2^2. For $3 \times 3 \times 3 \times 3$, we can write 3^4.
- In algebra, a variable is a letter used to stand for a number. The most common letters are: x, y, z, a, b, c, m, and n.
- Exponent's rules: $x^a \times x^b = x^{a+b}$, $\frac{x^a}{x^b} = x^{a-b}$.

$(x^a)^b = x^{a \times b}$ $\quad$ $(xy)^a = x^a \times y^a$ $\quad$ $\left(\frac{a}{b}\right)^c = \frac{a^c}{b^c}$

Examples:

Example 1. Multiply. $2x^2 \times 3x^4$

Solution: Use Exponent's rules: $x^a \times x^b = x^{a+b} \rightarrow x^2 \times x^4 = x^{2+4} = x^6$.
Then: $2x^2 \times 3x^4 = 6x^6$.

Example 2. Simplify. $(x^4y^2)^2$

Solution: Use Exponent's rules: $(x^a)^b = x^{a \times b}$.
Then: $(x^4y^2)^2 = x^{4\times2}y^{2\times2} = x^8y^4$.

Example 3. Multiply. $5x^8 \times 6x^5$

Solution: Use Exponent's rules: $x^a \times x^b = x^{a+b} \rightarrow x^8 \times x^5 = x^{8+5} = x^{13}$.
Then: $5x^8 \times 6x^5 = 30x^{13}$.

Example 4. Simplify. $(x^4y^7)^3$

Solution: Use Exponent's rules: $(x^a)^b = x^{a \times b}$.
Then: $(x^4y^7)^3 = x^{4\times3}y^{7\times3} = x^{12}y^{21}$.

Example 5. Simplify. $(x^3y^5)^2$

Solution: Use Exponent's rules: $(x^a)^b = x^{a \times b}$.
Then: $(x^3y^5)^2 = x^{3\times2}y^{5\times2} = x^6y^{10}$.

Division Property of Exponents

For division of exponents use following formulas:

- $\frac{x^a}{x^b} = x^{a-b}$ $(x \neq 0)$
- $\frac{x^a}{x^b} = \frac{1}{x^{b-a}}$ $(x \neq 0)$
- $\frac{1}{x^b} = x^{-b}$

Examples:

Example 1. Simplify. $\frac{16x^3y}{2xy^2} =$

Solution: First, cancel the common factor: $2 \rightarrow \frac{16x^3y}{2xy^2} = \frac{8x^3y}{xy^2}$.

Use Exponent's rules: $\frac{x^a}{x^b} = x^{a-b} \rightarrow \frac{x^3}{x} = x^{3-1} = x^2$ and $\frac{x^a}{x^b} = \frac{1}{x^{b-a}} \rightarrow \frac{y}{y^2} = \frac{1}{y^{2-1}} = \frac{1}{y}$.

Then: $\frac{16x^3y}{2xy^2} = \frac{8x^2}{y}$.

Example 2. Simplify. $\frac{24x^8}{3x^6} =$

Solution: Use Exponent's rules: $\frac{x^a}{x^b} = x^{a-b} \rightarrow \frac{x^8}{x^6} = x^{8-6} = x^2$.

Then: $\frac{24x^8}{3x^6} = 8x^2$.

Example 3. Simplify. $\frac{7x^4y^2}{28x^3y} =$

Solution: First, cancel the common factor: $7 \rightarrow \frac{x^4y^2}{4x^3y}$.

Use Exponent's rules: $\frac{x^a}{x^b} = x^{a-b} \rightarrow \frac{x^4}{x^3} = x^{4-3} = x$ and $\frac{y^2}{y} = y$.

Then: $\frac{7x^4y^2}{28x^3y} = \frac{xy}{4}$.

Example 4. Simplify. $\frac{8x^3y}{40x^2y^3} =$

Solution: First, cancel the common factor: $8 \rightarrow \frac{8x^3y}{40x^2y^3} = \frac{x^3y}{5x^2y^3}$.

Use Exponent's rules: $\frac{x^a}{x^b} = x^{a-b} \rightarrow \frac{x^3}{x^2} = x^{3-2} = x$.

Then: $\frac{8x^3y}{40x^2y^3} = \frac{xy}{5y^3} \rightarrow$ now cancel the common factor: $y \rightarrow \frac{xy}{5y^3} = \frac{x}{5y^2}$.

bit.ly/37JAclZ
Find more at

Powers of Products and Quotients

- For any non-zero numbers a and b and any integer x, $(ab)^x = a^x \times b^x$ and $\left(\frac{a}{b}\right)^c = \frac{a^c}{b^c}$.

Examples:

Example 1. Simplify. $(3x^3y^2)^2$

Solution: Use Exponent's rules: $(x^a)^b = x^{a\times b}$.

$$(3x^3y^2)^2 = (3)^2(x^3)^2(y^2)^2 = 9x^{3\times2}y^{2\times2} = 9x^6y^4.$$

Example 2. Simplify. $\left(\frac{2x^3}{3x^2}\right)^2$

Solution: First, cancel the common factor: $x \rightarrow \left(\frac{2x^3}{3x^2}\right) = \left(\frac{2x}{3}\right)^2$.

Use Exponent's rules: $\left(\frac{a}{b}\right)^c = \frac{a^c}{b^c}$. Then:

$$\left(\frac{2x}{3}\right)^2 = \frac{(2x)^2}{(3)^2} = \frac{4x^2}{9}.$$

Example 3. Simplify. $(6x^2y^4)^2$

Solution: Use Exponent's rules: $(x^a)^b = x^{a\times b}$.

$$(6x^2y^4)^2 = (6)^2(x^2)^2(y^4)^2 = 36x^{2\times2}y^{4\times2} = 36x^4y^8.$$

Example 4. Simplify. $(-4x^3y^5)^2$

Solution: Use Exponent's rules: $(x^a)^b = x^{a\times b}$.

$(-4x^3y^5)^2 = (-4)^2(x^3)^2(y^5)^2 = 16x^{3\times2}y^{5\times2} = 16x^6y^{10}$.

Example 5. Simplify. $\left(\frac{5x}{4x^2}\right)^2$

Solution: First, cancel the common factor: $x \rightarrow \left(\frac{5x}{4x^2}\right)^2 = \left(\frac{5}{4x}\right)^2$.

Use Exponent's rules: $\left(\frac{a}{b}\right)^c = \frac{a^c}{b^c}$. Then:

$$\left(\frac{5}{4x}\right)^2 = \frac{5^2}{(4x)^2} = \frac{25}{16x^2}.$$

Zero and Negative Exponents

- Zero-Exponent Rule: $a^0 = 1$, this means that anything raised to the zero power is 1. For example: $(5xy)^0 = 1$. (Number zero is an exception: $0^0 = 0$.)
- A negative exponent simply means that the base is on the wrong side of the fraction line, so you need to flip the base to the other side. For instance, "x^{-2}" (pronounced as "ecks to the minus two") just means "x^2" but underneath, as in $\frac{1}{x^2}$.

Examples:

Example 1. Evaluate. $\left(\frac{4}{5}\right)^{-2} =$

Solution: Use negative exponent's rule:

$$\left(\frac{x^a}{x^b}\right)^{-2} = \left(\frac{x^b}{x^a}\right)^2 \rightarrow \left(\frac{4}{5}\right)^{-2} = \left(\frac{5}{4}\right)^2.$$

Then: $\left(\frac{5}{4}\right)^2 = \frac{5^2}{4^2} = \frac{25}{16}$.

Example 2. Evaluate. $\left(\frac{3}{2}\right)^{-3} =$

Solution: Use negative exponent's rule:

$$\left(\frac{x^a}{x^b}\right)^{-3} = \left(\frac{x^b}{x^a}\right)^3 \rightarrow \left(\frac{3}{2}\right)^{-3} = \left(\frac{2}{3}\right)^3 =.$$

Then: $\left(\frac{2}{3}\right)^3 = \frac{2^3}{3^3} = \frac{8}{27}$.

Example 3. Evaluate. $\left(\frac{a}{b}\right)^0 =$

Solution: Use zero-exponent rule: $a^0 = 1$.

Then: $\left(\frac{a}{b}\right)^0 = 1$.

Example 4. Evaluate. $\left(\frac{4}{7}\right)^{-1} =$

Solution: Use negative exponent's rule:

$$\left(\frac{x^a}{x^b}\right)^{-1} = \left(\frac{x^b}{x^a}\right)^1 \rightarrow \left(\frac{4}{7}\right)^{-1} = \left(\frac{7}{4}\right)^1 = \frac{7}{4}.$$

bit.ly/3rnkh4v

Find more at

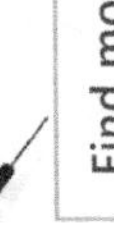

Negative Exponents and Negative Bases

- A negative exponent is the reciprocal of that number with a positive exponent. $(3)^{-2} = \frac{1}{3^2}$.
- To simplify a negative exponent, make the power positive!
- The parenthesis is important! -5^{-2} is not the same as $(-5)^{-2}$:

$$-5^{-2} = -\frac{1}{5^2} \text{ and } (-5)^{-2} = +\frac{1}{5^2}$$

Examples:

Example 1. Simplify. $\left(\frac{2a}{3c}\right)^{-2} =$

Solution: Use negative exponent's rule:

$$\left(\frac{x^a}{x^b}\right)^{-2} = \left(\frac{x^b}{x^a}\right)^2 \rightarrow \left(\frac{2a}{3c}\right)^{-2} = \left(\frac{3c}{2a}\right)^2.$$

Now, use the exponent's rule:

$$\left(\frac{a}{b}\right)^c = \frac{a^c}{b^c} \rightarrow \left(\frac{3c}{2a}\right)^2 = \frac{3^2c^2}{2^2a^2}.$$

Then: $\frac{3^2c^2}{2^2a^2} = \frac{9c^2}{4a^2}$.

Example 2. Simplify. $\left(\frac{x}{4y}\right)^{-3} =$

Solution: Use negative exponent's rule:

$$\left(\frac{x^a}{x^b}\right)^{-3} = \left(\frac{x^b}{x^a}\right)^3 \rightarrow \left(\frac{x}{4y}\right)^{-3} = \left(\frac{4y}{x}\right)^3.$$

Now, use the exponent's rule:

$$\left(\frac{a}{b}\right)^c = \frac{a^c}{b^c} \rightarrow \left(\frac{4y}{x}\right)^3 = \frac{4^3y^3}{x^3} = \frac{64y^3}{x^3}.$$

Example 3. Simplify. $\left(\frac{5a}{2c}\right)^{-2} =$

Solution: Use negative exponent's rule:

$$\left(\frac{x^a}{x^b}\right)^{-2} = \left(\frac{x^b}{x^a}\right)^2 \rightarrow \left(\frac{5a}{2c}\right)^{-2} = \left(\frac{2c}{5a}\right)^2.$$

Now, use the exponent's rule:

$$\left(\frac{a}{b}\right)^c = \frac{a^c}{b^c} \rightarrow= \left(\frac{2c}{5a}\right)^2 = \frac{2^2c^2}{5^2a^2}.$$

Then: $\frac{2^2c^2}{5^2a^2} = \frac{4c^2}{25a^2}$.

Scientific Notation

- Scientific notation is used to write very big or very small numbers in decimal form.
- In scientific notation, all numbers are written in the form of: $m \times 10^n$, where m is greater than 1 and less than 10.
- To convert a number from scientific notation to standard form, move the decimal point to the left (If the exponent of ten is a negative number), or to the right (If the exponent is positive).

Examples:

Example 1. Write 0.00024 in scientific notation.

Solution: First, move the decimal point to the right so you have a number between 1 and 10. That number is 2.4. Now, determine how many places the decimal moved in step 1 by the power of 10. We moved the decimal point 4 digits to the right. Then: $10^{-4} \rightarrow$ When the decimal is moved to the right, the exponent is negative. Then: $0.00024 = 2.4 \times 10^{-4}$.

Example 2. Write 3.8×10^{-5} in standard notation.

Solution: The exponent is negative 5. Then, move the decimal point to the left five digits. (Remember $3.8 = 0000003.8$.) When the decimal is moved to the right, the exponent is negative. Then: $3.8 \times 10^{-5} = 0.000038$.

Example 3. Write 0.00031 in scientific notation.

Solution: First, move the decimal point to the right so you have a number between 1 and 10. Then: $m = 3.1$. Now, determine how many places the decimal moved in step 1 by the power of 10. $10^{-4} \rightarrow$ Then: $0.00031 = 3.1 \times 10^{-4}$.

Example 4. Write 6.2×10^5 in standard notation.

Solution: $10^5 \rightarrow$ The exponent is positive 5.
Then, move the decimal point to the right five digits.
(Remember $6.2 = 6.20000$.) Then: $6.2 \times 10^5 = 620{,}000$.

Addition and Subtraction in Scientific Notation

- To add or subtract numbers in scientific notions, we need to have the same power as the base (number 10).
- Adding and subtracting numbers in scientific notion:
  - Step 1: Adjust the powers in the numbers so that they have the same power. (It is easier to adjust the smaller power to equal the larger one.)
  - Step 2: Add or subtract the numbers.
  - Step 3: Convert the answer to scientific notation if needed.

Examples:

Write the answers in scientific notation.

Example 1. $3.9 \times 10^5 + 4.2 \times 10^5$

Solution: Since two numbers have the same power, factor 10^5 out.

$(3.9 + 4.2) \times 10^5 = 8.1 \times 10^5$.

Example 2. $7.6 \times 10^9 - 5.5 \times 10^9$

Solution: Since two numbers have the same power, factor 10^9 out.

$(7.6 - 5.5) \times 10^9 = 2.1 \times 10^9$.

Example 3. $6.4 \times 10^7 - 3.2 \times 10^6$.

Solution: Convert the second number to have the same power of 10.

$3.2 \times 10^6 = 0.32 \times 10^7$.

Now, two numbers have the same power of 10.

Subtract: $6.4 \times 10^7 - 0.32 \times 10^7 = 6.08 \times 10^7$.

Multiplication and Division in Scientific Notation

- When multiplying two numbers in scientific notation, the process involves multiplying their coefficients and adding their exponents. This way, the product of the two numbers can be expressed in a concise form that is easier to work with and understand. The result of the multiplication will also be in scientific notation, allowing for efficient computation and manipulation of large or small numbers.
- To divide two numbers in scientific notation, divide their coefficients, and subtract their exponents.

Examples:

Write the answers in scientific notation.

Example 1. $(1.2 \times 10^5)(3 \times 10^{-2}) =$

Solution: First, multiply the coefficients: $1.2 \times 3 = 3.6$.
Add the powers of 10: $10^5 \times 10^{-2} = 10^3$.
Then: $(1.2 \times 10^5)(3 \times 10^{-2}) = 3.6 \times 10^3$.

Example 2. $\frac{2.8\times10^{-4}}{7\times10^6} =$

Solution: First, divide the coefficients: $\frac{2.8}{7} = 0.4$
Subtract the power of the exponent in the denominator from the exponent in the numerator: $\frac{10^{-4}}{10^6} = 10^{-4-6} = 10^{-10}$. Then: $\frac{2.8\times10^{-4}}{7\times10^6} = 0.4 \times 10^{-10}$.
Now, convert the answer to scientific notation: $0.4 \times 10^{-10} = 4 \times 10^{-11}$.

Example 3. $(4.3 \times 10^7)(5 \times 10^9) =$

Solution: First, multiply the coefficients: $4.3 \times 5 = 21.5$
Add the powers of 10: $10^7 \times 10^9 = 10^{16}$. Then: $(4.3 \times 10^7)(5 \times 10^9) = 21.5 \times 10^{16}$.
Now, convert the answer to scientific notation: $21.5 \times 10^{16} =$
2.15×10^{17}.

Chapter 2: Practices

Find the products.

1) $x^2 \times 4xy^2 =$
2) $3x^2y \times 5x^3y^2 =$
3) $6x^4y^2 \times x^2y^3 =$
4) $7xy^3 \times 2x^2y =$
5) $-5x^5y^5 \times x^3y^2 =$
6) $-8x^3y^2 \times 3x^3y^2 =$
7) $-6x^2y^6 \times 5x^4y^2 =$
8) $-3x^3y^3 \times 2x^3y^2 =$
9) $-6x^5y^3 \times 4x^4y^3 =$
10) $-2x^4y^3 \times 5x^6y^2 =$
11) $-7y^6 \times 3x^6y^3 =$
12) $-9x^4 \times 2x^4y^2 =$

Simplify.

13) $\frac{5^3 \times 5^4}{5^9 \times 5} =$
14) $\frac{3^3 \times 3^2}{7^2 \times 7} =$
15) $\frac{15x^5}{5x^3} =$
16) $\frac{16x^3}{4x^5} =$
17) $\frac{72y^2}{8x^3y^6} =$
18) $\frac{10x^3y^4}{50x^2y^3} =$
19) $\frac{13y^2}{52x^4y^4} =$
20) $\frac{50xy^3}{200x^3y^4} =$
21) $\frac{48x^2}{56x^2y^2} =$
22) $\frac{81y^6x}{54x^4y^3} =$

Solve.

23) $(x^3y^3)^2 =$
24) $(3x^3y^4)^3 =$
25) $(4x \times 6xy^3)^2 =$
26) $(5x \times 2y^3)^3 =$
27) $\left(\frac{9x}{x^3}\right)^2 =$
28) $(\frac{3y}{18y^2})^2 =$
29) $\left(\frac{3x^2y^3}{24x^4y^2}\right)^3 =$
30) $\left(\frac{26x^5y^3}{52x^3y^5}\right)^2 =$
31) $\left(\frac{18x^7y^4}{72x^5y^2}\right)^2 =$
32) $\left(\frac{12x^6y^4}{48x^5y^3}\right)^2 =$

Evaluate each expression.

33) $(\frac{1}{4})^{-2} =$

34) $(\frac{1}{3})^{-2} =$

35) $(\frac{1}{7})^{-3} =$

36) $(\frac{2}{5})^{-3} =$

37) $(\frac{2}{3})^{-3} =$

38) $(\frac{3}{5})^{-4} =$

Write each expression with positive exponents.

39) $x^{-7} =$

40) $3y^{-5} =$

41) $15y^{-3} =$

42) $-20x^{-4} =$

43) $12a^{-3}b^{5} =$

44) $25a^{3}b^{-4}c^{-3} =$

45) $-4x^{5}y^{-3}z^{-6} =$

46) $\frac{18y}{x^{3}y^{-2}} =$

47) $\frac{20a^{-2}b}{-12c^{-4}} =$

Write each number in scientific notation.

48) $0.00412 =$

49) $0.000053 =$

50) $66{,}000 =$

51) $72{,}000{,}000 =$

Write the answer in scientific notation.

52) $6 \times 10^{4} + 10 \times 10^{4} =$ ________

53) $7.2 \times 10^{6} - 3.3 \times 10^{6} =$ __________

54) $2.23 \times 10^{7} + 5.2 \times 10^{7} =$ ______

55) $8.3 \times 10^{9} - 5.6 \times 10^{8} =$ _____

56) $1.4 \times 10^{2} + 7.4 \times 10^{5} =$ _____

57) $9.6 \times 10^{6} - 3 \times 10^{4} =$ ______

Simplify. Write the answer in scientific notation.

58) $(5.6 \times 10^{12})(3 \times 10^{-7}) =$ ____

59) $(3 \times 10^{-8})(7 \times 10^{10}) =$ _____

60) $(9 \times 10^{-3})(4.2 \times 10^{6}) =$ ____

61) $\frac{125\times10^{9}}{50\times10^{12}} =$ ____

62) $\frac{2.8\times10^{12}}{0.4\times10^{20}} =$ ____

63) $\frac{9\times10^{8}}{3\times10^{7}} =$ ____

Chapter 2: Answers

1) $4x^3y^2$
2) $15x^5y^3$
3) $6x^6y^5$
4) $14x^3y^4$
5) $-5x^8y^7$
6) $-24x^6y^4$
7) $-30x^6y^8$
8) $-6x^6y^5$
9) $-24x^9y^6$
10) $-10x^{10}y^5$
11) $-21x^6y^9$
12) $-18x^8y^2$
13) $\frac{1}{125}$
14) $\frac{243}{343}$
15) $3x^2$
16) $\frac{4}{x^2}$
17) $\frac{9}{x^3y^4}$
18) $\frac{xy}{5}$
19) $\frac{1}{4x^4y^2}$
20) $\frac{1}{4x^2y}$
21) $\frac{6}{7y^2}$
22) $\frac{3y^3}{2x^3}$
23) x^6y^6
24) $27x^9y^{12}$
25) $576x^4y^6$
26) $1{,}000x^3y^9$
27) $\frac{81}{x^4}$
28) $\frac{1}{36y^2}$
29) $\frac{y^3}{512x^6}$
30) $\frac{x^4}{4y^4}$
31) $\frac{x^4y^4}{16}$
32) $\frac{x^2y^2}{16}$
33) 16
34) 9
35) 343
36) $\frac{125}{8}$
37) $\frac{27}{8}$
38) $\frac{625}{81}$
39) $\frac{1}{x^7}$
40) $\frac{3}{y^5}$
41) $\frac{15}{y^3}$
42) $-\frac{20}{x^4}$
43) $\frac{12b^5}{a^3}$
44) $\frac{25a^3}{b^4c^3}$
45) $-\frac{4x^5}{y^3z^6}$
46) $\frac{18y^3}{x^3}$
47) $-\frac{5bc^4}{3a^2}$
48) 4.12×10^{-3}
49) 5.3×10^{-5}
50) 6.6×10^4
51) 7.2×10^7
52) 1.6×10^5

53) 3.9×10^6
54) 7.43×10^7
55) 7.74×10^9
56) 7.4014×10^5
57) 9.57×10^6
58) 1.68×10^6
59) 2.1×10^3
60) 3.78×10^4
61) 2.5×10^{-3}
62) 7×10^{-8}
63) 3×10^1

CHAPTER

3 Expressions and Equations

Math topics that you'll learn in this chapter:

- ☑ Simplifying Variable Expressions
- ☑ Evaluating One Variable
- ☑ Evaluating Two Variables
- ☑ One–Step Equations
- ☑ Multi–Step Equations
- ☑ Rearrange Multi-Variable Equations
- ☑ Finding Midpoint
- ☑ Finding the Distance between Two Points

Simplifying Variable Expressions

- In algebra, a variable is a letter used to stand for a number. The most common letters are x, y, z, a, b, c, m, and n.
- An algebraic expression is an expression that contains integers, variables, and math operations such as addition, subtraction, multiplication, division, etc.
- In an expression, we can combine "like" terms. (Values with same variable and same power.)

Examples:

Example 1. Simplify. $(4x + 2x + 4) =$

Solution: In this expression, there are three terms: $4x$, $2x$, and 4. Two terms are "like" terms: $4x$ and $2x$. Combine like terms. $4x + 2x = 6x$.
Then:

$$(4x + 2x + 4) = 6x + 4.$$

(*Remember you cannot combine variables and numbers.*)

Example 2. Simplify. $-2x^2 - 5x + 4x^2 - 9 =$

Solution: Combine "like" terms: $-2x^2 + 4x^2 = 2x^2$.
Then:

$$-2x^2 - 5x + 4x^2 - 9 = 2x^2 - 5x - 9.$$

Example 3. Simplify. $(-8 + 6x^2 + 3x^2 + 9x) =$

Solution: Combine "like" terms: $6x^2 + 3x^2 = 9x^2$.
Then:

$$(-8 + 6x^2 + 3x^2 + 9x) = 9x^2 + 9x - 8.$$

Example 4. Simplify. $-10x + 6x^2 - 3x + 9x^2 =$

Solution: Combine "like" terms: $-10x - 3x = -13x$, and $6x^2 + 9x^2 = 15x^2$.
Then:

$$-10x + 6x^2 - 3x + 9x^2 = -13x + 15x^2.$$

Write in standard form (biggest powers first):
$-13x + 15x^2 = 15x^2 - 13x$.

Evaluating One Variable

- To evaluate one-variable expression, find the variable and substitute a number for that variable.
- Perform the arithmetic operations.

Examples:

Example 1. Calculate this expression for $x = 2$: $8 + 2x$.

Solution: First, substitute 2 for x.
Then: $8 + 2x = 8 + 2(2)$.
Now, use the order of operation to find the answer:

$$8 + 2(2) = 8 + 4 = 12.$$

Example 2. Evaluate this expression for $x = -1$: $4x - 8$.

Solution: First, substitute -1 for x.
Then: $4x - 8 = 4(-1) - 8$.
Now, use the order of operation to find the answer:

$$4(-1) - 8 = -4 - 8 = -12.$$

Example 3. Find the value of this expression when $x = 4$. $(16 - 5x)$.

Solution: First, substitute 4 for x,
then:

$$16 - 5x = 16 - 5(4) = 16 - 20 = -4.$$

Example 4. Solve this expression for $x = -3$: $15 + 7x$.

Solution: Substitute -3 for x.
Then:

$$15 + 7x = 15 + 7(-3) = 15 - 21 = -6.$$

Example 5. Solve this expression for $x = -2$: $12 + 3x$.

Solution: Substitute -2 for x,

then: $12 + 3x = 12 + 3(-2) = 12 - 6 = 6$.

Evaluating Two Variables

- To evaluate an algebraic expression, substitute a number for each variable.
- Evaluating an algebraic expression involves replacing variables in the expression with specific numerical values to obtain a single numerical value as the answer. To do this, the following steps can be followed:

  1- Substitute the given numerical values for the variables in the expression.

  2- Simplify the expression by performing the necessary arithmetic operations (addition, subtraction, multiplication, division, exponentiation, etc.) in the order prescribed by the rules of arithmetic.

  3- Continue simplifying the expression until you obtain a single numerical value as the answer.

Examples:

Example 1. Calculate this expression for $a = 2$ and $b = -1$: $(4a - 3b)$.

Solution: First, substitute 2 for a, and -1 for b.
Then: $4a - 3b = 4(2) - 3(-1)$.
Now, use the order of operation to find the answer:

$$4(2) - 3(-1) = 8 + 3 = 11.$$

Example 2. Evaluate this expression for $x = -2$ and $y = 2$: $(3x + 6y)$.

Solution: Substitute -2 for x, and 2 for y.
Then:

$$3x + 6y = 3(-2) + 6(2) = -6 + 12 = 6.$$

Example 3. Find the value of this expression $2(6a - 5b)$, when $a = -1$ and $b = 4$.

Solution: Substitute -1 for a, and 4 for b.
Then:

$$2(6a - 5b) = 2\big(6(-1) - 5(4)\big) = 2(-6 - 20) = 2(-26) = -52.$$

Example 4. Evaluate this expression. $-7x - 2y$, $x = 4$, $y = -3$.

Solution: Substitute 4 for x, and -3 for y and simplify.
Then: $-7x - 2y = -7(4) - 2(-3) = -28 + 6 = -22$.

One–Step Equations

- The values of two expressions on both sides of an equation are equal. Example: $ax = b$. In this equation, ax is equal to b.
- Solving an equation means finding the value of the variable.
- You only need to perform one Math operation to solve the one-step equations.
- To solve a one-step equation, find the inverse (opposite) operation that is being performed.
- The inverse operations are:
  - ❖ Addition and subtraction
  - ❖ Multiplication and division

Examples:

Example 1. Solve this equation for x: $4x = 16 \rightarrow x = ?$

Solution: Here, the operation is multiplication (Variable x is multiplied by 4.) and its inverse operation is division. To solve this equation, divide both sides of equation by 4:

$$4x = 16 \rightarrow \frac{4x}{4} = \frac{16}{4} \rightarrow x = 4.$$

Example 2. Solve this equation: $x + 8 = 0 \rightarrow x = ?$

Solution: In this equation, 8 is added to the variable x. The inverse operation of addition is subtraction. To solve this equation, subtract 8 from both sides of the equation:

$$x + 8 - 8 = 0 - 8.$$

Then: $x = -8$.

Example 3. Solve this equation for x. $x - 12 = 0$

Solution: Here, the operation is subtraction, and its inverse operation is addition. To solve this equation, add 12 to both sides of the equation:

$$x - 12 + 12 = 0 + 12 \rightarrow x = 12.$$

Multi–Step Equations

- To solve a multi-step equation, combine "like" terms on one side.
- Bring variables to one side by adding or subtracting.
- Simplify using the inverse of addition or subtraction.
- Simplify further by using the inverse of multiplication or division.
- Check your solution by plugging the value of the variable into the original equation.

Examples:

Example 1. Solve this equation for x. $4x + 8 = 20 - 2x$

Solution: First, bring variables to one side by adding $2x$ to both sides. Then:

$$4x + 8 + 2x = 20 - 2x + 2x \rightarrow 4x + 8 + 2x = 20.$$

Simplify: $6x + 8 = 20$. Now, subtract 8 from both sides of the equation:

$6x + 8 - 8 = 20 - 8 \rightarrow 6x = 12 \rightarrow$ Divide both sides by 6:

$$6x = 12 \rightarrow \frac{6x}{6} = \frac{12}{6} \rightarrow x = 2.$$

Let's check this solution by substituting the value of 2 for x in the original equation:

$$x = 2 \rightarrow 4x + 8 = 20 - 2x \rightarrow 4(2) + 8 = 20 - 2(2) \rightarrow 16 = 16.$$

The answer $x = 2$ is correct.

Example 2. Solve this equation for x. $-5x + 4 = 24$

Solution: Subtract 4 from both sides of the equation.

$$-5x + 4 = 24 \rightarrow -5x + 4 - 4 = 24 - 4 \rightarrow -5x = 20.$$

Divide both sides by -5, then:

$$-5x = 20 \rightarrow \frac{-5x}{-5} = \frac{20}{-5} \rightarrow x = -4.$$

Now, check the solution:

$$x = -4 \rightarrow -5x + 4 = 24 \rightarrow -5(-4) + 4 = 24 \rightarrow 24 = 24.$$

The answer $x = -4$ is correct.

Rearrange Multi-Variable Equations

- In order to rearrange the multi-variable equations for each of the variables:
  - Determine the dependent variable.
  - Isolate the dependent variable on both sides of the equation.
  - By undoing the operations on both sides, write the dependent variable on one side and the other variables on the other side of the equation.

Examples:

Example 1. Make x as the subject in the equation: $\frac{1}{2}x - t = 5$.

Solution: Find x like this $\frac{1}{2}x - t = 5$. To isolate x on one side of the equation, first add t to both sides, as follows: $\frac{1}{2}x - t + t = 5 + t \rightarrow \frac{1}{2}x = t + 5$. Then multiply 2 on both sides of the equation: $2\left(\frac{1}{2}x\right) = 2(t + 5) \rightarrow x = 2t + 10$.

Example 2. Solve for a in terms of b and c: $-c + b - a = 6$.

Solution: To solve the problem, find a. By undoing the operations on both sides, isolate a on one side of the equation. Add a to both sides of the equation. So,

$$-c + b - a + a = 6 + a \rightarrow -c + b = a + 6.$$

Now, subtract 6 from both sides of the equation, then:

$$-c + b - 6 = a + 6 - 6 \rightarrow -c + b - 6 = a.$$

In this case, the above equation in terms of a becomes: $a = -c + b - 6$

Example 3. Solve $V = \frac{1}{3}\pi r^2 h$ for h.

Solution: To isolate h on both sides of the equation, just multiply the sides by the expression $\frac{3}{\pi r^2}$. Therefore, $\frac{3}{\pi r^2} \times V = \frac{3}{\pi r^2}\left(\frac{1}{3}\pi r^2 h\right) \rightarrow h = \frac{3V}{\pi r^2}$.

Example 4. Solve $y = mx + c$ for m.

Solution: Specify m as the dependent variable like this $y = mx + c$. Subtract c from both sides of the equation: $y - c = mx + c - c \rightarrow y - c = mx$. In this case, divide both sides of the obtained equation by x, so, we have:

$$\frac{y-c}{x} = \frac{mx}{x} \rightarrow m = \frac{y-c}{x}.$$

Finding Midpoint

- The middle of a line segment is its midpoint.
- The midpoint of two endpoints $A(x_1, y_1)$ and $B(x_2, y_2)$ can be found using this formula: $M\left(\frac{x_1+x_2}{2}, \frac{y_1+y_2}{2}\right)$.

Examples:

Example 1. Find the midpoint of the line segment with the given endpoints. $(2,-4)$, $(6,8)$

Solution: Midpoint $= \left(\frac{x_1+x_2}{2}, \frac{y_1+y_2}{2}\right) \rightarrow (x_1, y_1) = (2,-4)$ and $(x_2, y_2) = (6,8)$.

Midpoint $= \left(\frac{2+6}{2}, \frac{-4+8}{2}\right) \rightarrow \left(\frac{8}{2}, \frac{4}{2}\right) \rightarrow M(4,2)$.

Example 2. Find the midpoint of the line segment with the given endpoints. $(-2,3)$, $(6,-7)$

Solution: Midpoint $= \left(\frac{x_1+x_2}{2}, \frac{y_1+y_2}{2}\right) \rightarrow (x_1, y_1) = (-2,3)$ and $(x_2, y_2) = (6,-7)$.

Midpoint $= \left(\frac{-2+6}{2}, \frac{3+(-7)}{2}\right) \rightarrow \left(\frac{4}{2}, \frac{-4}{2}\right) \rightarrow M(2,-2)$.

Example 3. Find the midpoint of the line segment with the given endpoints. $(7,-4)$, $(1,8)$

Solution: Midpoint $= \left(\frac{x_1+x_2}{2}, \frac{y_1+y_2}{2}\right) \rightarrow (x_1, y_1) = (7,-4)$ and $(x_2, y_2) = (1,8)$.

Midpoint $= \left(\frac{7+1}{2}, \frac{-4+8}{2}\right) \rightarrow \left(\frac{8}{2}, \frac{4}{2}\right) \rightarrow M(4,2)$.

Example 4. Find the midpoint of the line segment with the given endpoints. $(6,3)$, $(10,-9)$

Solution: Midpoint $= \left(\frac{x_1+x_2}{2}, \frac{y_1+y_2}{2}\right) \rightarrow (x_1, y_1) = (6,3)$ and $(x_2, y_2) = (10,-9)$.

Midpoint $= \left(\frac{6+10}{2}, \frac{3-9}{2}\right) \rightarrow \left(\frac{16}{2}, \frac{-6}{2}\right) \rightarrow M(8,-3)$.

Finding the Distance between Two Points

- Use the following formula to find the distance of two points with the coordinates $A(x_1, y_1)$ and $B(x_2, y_2)$:

$$d = \sqrt{(x_2 - x_1)^2 + (y_2 - y_1)^2}$$

Examples:

Example 1. Find the distance between $(4,2)$ and $(-5,-10)$ on the coordinate plane.

Solution: Use the distance of two points formula:

$$d = \sqrt{(x_2 - x_1)^2 + (y_2 - y_1)^2}.$$

Considering that: $(x_1, y_1) = (4,2)$ and $(x_2, y_2) = (-5,-10)$.

Then:

$$d = \sqrt{(-5-4)^2 + (-10-2)^2} = \sqrt{(-9)^2 + (-12)^2} = \sqrt{81 + 144} = \sqrt{225} = 15.$$

Then: $d = 15$.

Example 2. Find the distance of two points $(-1,5)$ and $(-4,1)$.

Solution: Use the distance of two points formula:

$$d = \sqrt{(x_2 - x_1)^2 + (y_2 - y_1)^2}.$$

Since $(x_1, y_1) = (-1,5)$, and $(x_2, y_2) = (-4,1)$.

Then:

$$d = \sqrt{(-4-(-1))^2 + (1-5)^2} = \sqrt{(-3)^2 + (-4)^2} = \sqrt{9 + 16} = \sqrt{25} = 5.$$

Then: $d = 5$.

Example 3. Find the distance between $(-6,5)$ and $(-1,-7)$.

Solution: Use the distance of two points formula:

$$d = \sqrt{(x_2 - x_1)^2 + (y_2 - y_1)^2}.$$

According to: $(x_1, y_1) = (-6,5)$ and $(x_2, y_2) = (-1,-7)$.

Then:

$$d = \sqrt{\left(-1-(-6)\right)^2 + (-7-5)^2} = \sqrt{(5)^2 + (-12)^2} =$$

$$\sqrt{25 + 144} = \sqrt{169} = 13.$$

Chapter 3: Practices

Simplify each expression.

1) $(3 + 4x - 1) =$
2) $(-5 - 2x + 7) =$
3) $(12x - 5x - 4) =$
4) $(-16x + 24x - 9) =$
5) $(6x + 5 - 15x) =$
6) $2 + 5x - 8x - 6 =$
7) $5x + 10 - 3x - 22 =$
8) $-5 - 3x^2 - 6 + 4x =$
9) $-6 + 9x^2 - 3 + x =$
10) $5x^2 + 3x - 10x - 3 =$
11) $4x^2 - 2x - 6x + 5 - 8 =$
12) $3x^2 - 5x - 7x + 2 - 4 =$
13) $9x^2 - x - 5x + 3 - 9 =$
14) $2x^2 - 7x - 3x^2 + 4x + 6 =$

Evaluate each expression using the value given.

15) $x = 4 \rightarrow 10 - x =$ ____
16) $x = 6 \rightarrow x + 8 =$ ____
17) $x = 3 \rightarrow 2x - 6 =$ ____
18) $x = 2 \rightarrow 10 - 4x =$ ____
19) $x = 7 \rightarrow 8x - 3 =$ ____
20) $x = 9 \rightarrow 20 - 2x =$ ____
21) $x = 5 \rightarrow 10x - 30 =$ ___
22) $x = -6 \rightarrow 5 - x =$ ____
23) $x = -3 \rightarrow 22 - 3x =$ ____
24) $x = -7 \rightarrow 10 - 9x =$ ____
25) $x = -10 \rightarrow 40 - 3x =$ ____
26) $x = -2 \rightarrow 20x - 5 =$ ____
27) $x = -5 \rightarrow -10x - 8 =$ ___
28) $x = -4 \rightarrow -1 - 4x =$ ___

Evaluate each expression using the values given.

29) $x = 2, y = 1 \rightarrow 2x + 7y =$ _____
30) $a = 3, b = 5 \rightarrow 3a - 5b =$ _____
31) $x = 6, y = 2 \rightarrow 3x - 2y + 8 =$ _____
32) $a = -2, b = 3 \rightarrow -5a + 2b + 6 =$ _____
33) $x = -4, y = -3 \rightarrow -4x + 10 - 8y =$ _____

Solve each equation.

34) $x + 6 = 3 \rightarrow x =$ ____

35) $5 = 11 - x \rightarrow x =$ ____

36) $-3 = 8 + x \rightarrow x =$ ____

37) $x - 2 = -7 \rightarrow x =$ ____

38) $-15 = x + 6 \rightarrow x =$ ____

39) $10 - x = -2 \rightarrow x =$ ____

40) $22 - x = -9 \rightarrow x =$ ____

41) $-4 + x = 28 \rightarrow x =$ ____

42) $11 - x = -7 \rightarrow x =$ ____

43) $35 - x = -7 \rightarrow x =$ ____

Solve each equation.

44) $4(x + 2) = 12 \rightarrow x =$ ____

45) $-6(6 - x) = 12 \rightarrow x =$ ____

46) $5 = -5(x + 2) \rightarrow x =$ ____

47) $-10 = 2(4 + x) \rightarrow x =$ ____

48) $4(x + 2) = -12 \rightarrow x =$ ____

49) $-6(3 + 2x) = 30 \rightarrow x =$ ____

50) $-3(4 - x) = 12 \rightarrow x =$ ____

51) $-4(6 - x) = 16 \rightarrow x =$ ____

Solve.

52) $q = 2l + 2w$ for w.

53) $x = 2yw$ for w.

54) $pv = nRT$ for T.

55) $a = b + c + d$ for d.

Find the midpoint of the line segment with the given endpoints.

56) $(5,0), (1,4)$

57) $(2,3), (4,7)$

58) $(8,1), (2,5)$

59) $(5,10), (3,6)$

60) $(4, -1), (-2,7)$

61) $(2, -5), (4,1)$

62) $(7, 6), (-5, 2)$

63) $(-2, 8), (4, -6)$

Find the distance between each pair of points.

64) $(-2,8), (-6,8)$

65) $(4,-4), (14,20)$

66) $(-1,9), (-5,6)$

67) $(0,3), (4,3)$

68) $(0,-2), (5,10)$

69) $(4,3), (7,-1)$

70) $(2,6), (10,-9)$

71) $(3,3), (6,-1)$

72) $(-2,-12), (14,18)$

73) $(2,-2), (12,22)$

Chapter 3: Answers

1) $4x + 2$
2) $-2x + 2$
3) $7x - 4$
4) $8x - 9$
5) $-9x + 5$
6) $-3x - 4$
7) $2x - 12$
8) $-3x^2 + 4x - 11$
9) $9x^2 + x - 9$
10) $5x^2 - 7x - 3$
11) $4x^2 - 8x - 3$
12) $3x^2 - 12x - 2$
13) $9x^2 - 6x - 6$
14) $-x^2 - 3x + 6$
15) 6
16) 14
17) 0
18) 2
19) 53
20) 2
21) 20
22) 11
23) 31
24) 73
25) 70
26) -45
27) 42
28) 15
29) 11
30) -16
31) 22
32) 22
33) 50
34) -3
35) 6
36) -11
37) -5
38) -21
39) 12
40) 31
41) 32
42) 18
43) 42
44) 1
45) 8
46) -3
47) -9
48) -5
49) -4
50) 8
51) 10

52) $\frac{1}{2}q - l = w$

53) $w = \frac{x}{2y}$

54) $T = \frac{PV}{nR}$

55) $d = a - b - c$

56) $(3, 2)$

57) $(3, 5)$

58) $(5,3)$

59) $(4,8)$

60) $(1, 3)$

61) $(3, -2)$

62) $(1, 4)$

63) $(1, 1)$

64) 4

65) 26

66) 5

67) 4

68) 13

69) 5

70) 17

71) 5

72) 34

73) 26

CHAPTER

4 Linear Functions

Math topics that you'll learn in this chapter:

- ☑ Finding Slope
- ☑ Writing Linear Equations
- ☑ Graphing Linear Inequalities
- ☑ Write an Equation from a Graph
- ☑ Slope-intercept Form and Point-slope Form
- ☑ Write a Point-slope Form Equation from a Graph
- ☑ Find $x-$ and $y-$intercepts in the Standard Form of Equation
- ☑ Graph an Equation in the Standard Form
- ☑ Equations of Horizontal and Vertical Lines
- ☑ Graph a Horizontal or Vertical line
- ☑ Graph an Equation in Point-Slope Form
- ☑ Equation of Parallel or Perpendicular Lines
- ☑ Compare Linear Function's Graph and Linear Equations
- ☑ Graphing Absolute Value Equations
- ☑ Two-variable Linear Equation Word Problems

Finding Slope

- The slope of a line represents the direction of a line on the coordinate plane.
- A coordinate plane contains two perpendicular number lines. The horizontal line is x and the vertical line is y. The point at which the two axes intersect is called the origin. An ordered pair (x, y) shows the location of a point.
- A line on a coordinate plane can be drawn by connecting two points.
- To find the slope of a line, we need the equation of the line or two points on the line.
- The slope of a line with two points $A(x_1, y_1)$ and $B(x_2, y_2)$ can be found by using this formula: $\frac{y_2-y_1}{x_2-x_1} = \frac{rise}{run}$.
- The equation of a line is typically written as $y = mx + b$ where m is the slope and b is the y −intercept.

Examples:

Example 1. Find the slope of the line through these two points:

A$(1, -6)$ and $B(3, 2)$.

Solution: Slope $= \frac{y_2-y_1}{x_2-x_1}$. Let (x_1, y_1) be $A(1, -6)$ and (x_2, y_2) be $B(3,2)$.

(Remember, you can choose any point for (x_1, y_1) and (x_2, y_2)).

Then:

$$\text{Slope} = \frac{y_2-y_1}{x_2-x_1} = \frac{2-(-6)}{3-1} = \frac{8}{2} = 4.$$

The slope of the line through these two points is 4.

Example 2. Find the slope of the line with equation $y = -2x + 8$.

Solution: When the equation of a line is written in the form of $y = mx + b$, the slope is m.

In this line: $y = -2x + 8$, the slope is -2.

Writing Linear Equations

- The equation of a line in slope-intercept form: $y = mx + b$.
- To write the equation of a line, first identify the slope.
- Find the y −intercept. This can be done by substituting the slope and the coordinates of a point (x, y) on the line.

Examples:

Example 1. What is the equation of the line that passes through $(3, -4)$ and has a slope of 6?

Solution: The general slope-intercept form of the equation of a line is:

$y = mx + b$,

where m is the slope and b is the y −intercept.
By substitution of the given point and given slope:

$y = mx + b \rightarrow -4 = (6)(3) + b$.

So, $b = -4 - 18 = -22$, and the required equation of the line is: $y = 6x - 22$.

Example 2. Write the equation of the line through two points $A(3,1)$ and $B(-2,6)$.

Solution: First, find the slope: slope $= \frac{y_2 - y_1}{x_2 - x_1} = \frac{6-1}{-2-3} = \frac{5}{-5} = -1 \rightarrow m = -1$.
To find the value of b, use either point and plug in the values of x and y in the equation. The answer will be the same: $y = -x + b$. Let's check both points.
Then: $(3,1) \rightarrow y = mx + b \rightarrow 1 = -1(3) + b \rightarrow b = 4$.

$(-2,6) \rightarrow y = mx + b \rightarrow 6 = -1(-2) + b \rightarrow b = 4$.

The y −intercept of the line is 4. The equation of the line is: $y = -x + 4$.

Example 3. What is the equation of the line that passes through $(4, -1)$ and has a slope of 4?

Solution: The general slope-intercept form of the equation of a line is:

$y = mx + b$, where m is the slope and b is the y −intercept.

By substitution of the given point and given slope:

$y = mx + b \rightarrow -1 = (4)(4) + b$.

So, $b = -1 - 16 = -17$, and the equation of the line is: $y = 4x - 17$.

Graphing Linear Inequalities

- To graph a linear inequality, first draw a graph of the "equals" line.
- Use a dashed line for less than (<) and greater than (>) signs and a solid line for less than and equal to (≤) and greater than and equal to (≥).
- Choose a testing point. (It can be any point on both sides of the line.)
- Put the value of (x, y) of that point in the inequality. If that works, that part of the line is the solution. If the values don't work, then the other part of the line is the solution.

Example:

Sketch the graph of inequality: $y < 2x + 4$.

Solution: To draw the graph of $y < 2x + 4$, you first need to graph the line:

$y = 2x + 4$.

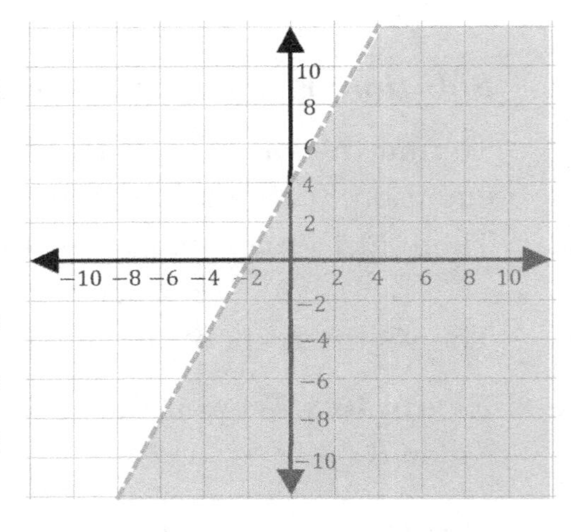

Since there is a less than (<) sign, draw a dashed line.

The slope is 2 and the y −intercept is 4.

Then, choose a testing point and substitute the value of x and y from that point into the inequality. The easiest point to test is the origin: (0,0).

$(0,0) \rightarrow y < 2x + 4 \rightarrow 0 < 2(0) + 4 \rightarrow 0 < 4.$

This is correct! 0 is less than 4. So, this part of the line (on the right side) is the solution of this inequality.

Write an Equation from a Graph

- To write an equation of a line in slope-intercept form, given a graph of that equation, pick two points on the line and use them to find the slope. This is the value of m in the equation. Next, find the coordinates of the y −intercept these should be of the form $(0, b)$. The y −coordinate is the value of b in the equation.
- Finally, write the equation, substituting numerical values for m and b. Check your equation by picking a point on the line (Not the y −intercept) and plugging it in to see if it satisfies the equation.

Example:

Write the equation of the following line in slope-intercept form.

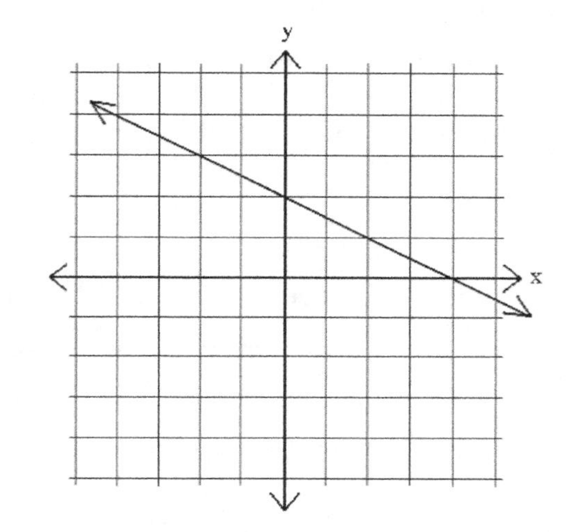

Solution: First, pick two points on the line for example, (2,1) and (4,0).

Use these points to calculate the slope:

$$m = \frac{0-1}{4-2} = \frac{-1}{2} = -\frac{1}{2}.$$

Next, find the y −intercept:

(0,2) Thus, $b = 2$.

Therefore, the equation of this line is:

$$y = -\frac{1}{2}x + 2.$$

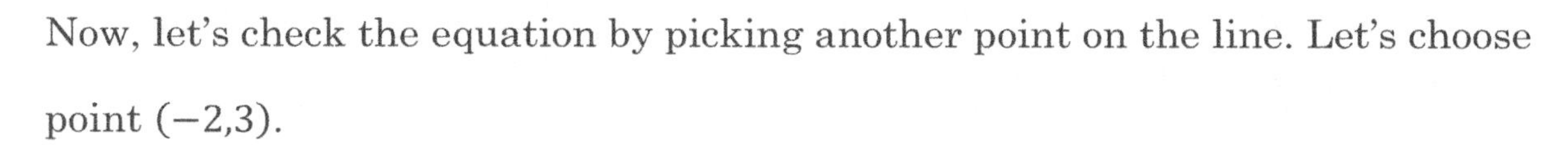

Now, let's check the equation by picking another point on the line. Let's choose point $(-2,3)$.

Then:

$$(-2,3) \rightarrow y = -\frac{1}{2}x + 2 \rightarrow 3 = -\frac{1}{2}(-2) + 2 \rightarrow 3 = 1 + 2.$$ This is true!

Slope-intercept Form and Point-slope Form

- The point-slope form of the equation of a straight line is: $y - y_1 = m(x - x_1)$. The equation is useful when we know: one point on the line: (x_1, y_1) and the slope of the line: m.
- The slope-intercept form is probably the most frequently used way to express the equation of a line. In general, the slope intercept form is:

$$y = mx + b$$

Examples:

Example 1. Find the equation of a line with point (1,5) and slope -3, and write it in slope-intercept and point-slope forms.

Solution: For point-slope form, we have the point and slope:

$$x_1 = 1,\ y_1 = 5,\ m = -3$$

Then: $y - y_1 = m(x - x_1) \rightarrow y - 5 = -3(x - 1)$

The slope-intercept form of a line is: $y = mx + b$

Since $y = 5, x = 1, m = -3$, we just need to solve for b

$$y = mx + b \rightarrow 5 = -3(1) + b \rightarrow b = 8$$

Slope-intercept form: $y = -3x + 8$

Example 2. Find the equation of a line with point (4,6) and slope 3, and write it in slope-intercept and point-slope forms.

Solution: For point-slope form, we have the point and slope:

$$x_1 = 4,\ y_1 = 6,\ m = 3$$

Then: $y - y_1 = m(x - x_1) \rightarrow y - 6 = 3(x - 4)$

The slope-intercept form of a line is: $y = mx + b$

Since $y = 6,\ x = 4,\ m = 3$, we just need to solve for b

$$y = mx + b \rightarrow 6 = 3(4) + b \rightarrow b = -6$$

Slope-intercept form: $y = 3x - 6$

Write a Point-slope Form Equation from a Graph

- The usage of the point-slope form equation is to determine the equation of a straight line when passes through a given point. In fact, you can use the point-slope formula only when you have the line's slope and a given point on the line. A line's equation with the slope of $'m'$ that passes through the point (x_1, y_1) can be found by the point-slope formula.
- The point-slope form equation is $y - y_1 = m(x - x_1)$. In this equation, (x_1, y_1) is considered as a random point on the line and m is a sign to represent the line's slope.
- To find the point-slope form equation of a straight line and solve it, you can follow the following steps:
  - 1st step: Find the slope, 'm' of the straight line. the slope formula is $m = \frac{change\ in\ y}{change\ in\ x}$. Then find the coordinates (x_1, y_1) of the random point on the line.
  - 2nd step: Put the values you found in the first step in the point-slope formula: $y - y_1 = m(x - x_1)$
  - 3rd step: Simplify the given equation to get the line's equation in the standard form.

Example:

According to the following graph, what is the equation of the line in point-slope form?

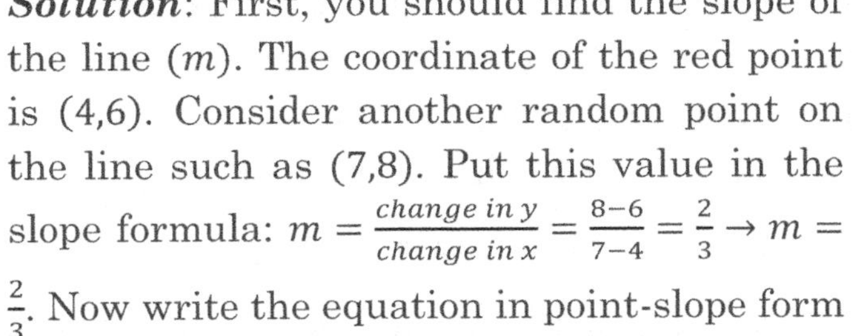

Solution: First, you should find the slope of the line (m). The coordinate of the red point is (4,6). Consider another random point on the line such as (7,8). Put this value in the slope formula: $m = \frac{change\ in\ y}{change\ in\ x} = \frac{8-6}{7-4} = \frac{2}{3} \rightarrow m = \frac{2}{3}$. Now write the equation in point-slope form using the coordinate of the red point is (4,6) and $m = \frac{2}{3}$: $y - y_1 = m(x - x_1) \rightarrow y - 6 = \frac{2}{3}(x - 4)$. Therefore, the equation of the line in point-slope form is $y - 6 = \frac{2}{3}(x - 4)$.

bit.ly/3yaeJ1Q

Find more at

Find $x-$ and $y-$intercepts in the Standard Form of Equation

- The linear equations' standard form (the general form) is described as $Ax + By = C$. In this form of the equation, A, B, and C are integers, and the letters x and y are considered the variables.
- When you need to have a linear equation in the standard form, you can easily change it in a way that can be represented in the form $Ax + By = C$. Keep in mind that A, B, and C should be integers and the order of the variables should be as mentioned in the standard form's equation.
- A line's $x-$intercept is the $x-$value when the line intersects the $x-$axis. In this case, y is equal to zero.
- A line's $y-$intercept is the $y-$value where the line intersects the $y-$axis. In this case, x is equal to zero.

Examples:

Example 1. Find the $x-$ and $y-$intercepts of line $6x + 24y = 12$.

Solution: To find the $x-$intercept, you can consider y equal to 0 and solve for x: $6x + 24y = 12 \rightarrow 6x + 24(0) = 12 \rightarrow 6x = 12 \rightarrow x = 2$. The $x-$intercept is 2. To find the $y-$intercept, you can consider x equal to 0 and solve for y: $6x + 24y = 12 \rightarrow 6(0) + 24y = 12 \rightarrow 24y = 12 \rightarrow y = \frac{1}{2}$. The $y-$intercept is $\frac{1}{2}$.

Example 2. Find the $x-$ and $y-$intercepts of line $3x + 5y = -15$.

Solution: To find the $x-$intercept, you can consider y equal to 0 and solve for x: $3x + 5y = -15 \rightarrow 3x + 5(0) = -15 \rightarrow 3x = -15 \rightarrow x = -5$. The $x-$intercept is -5. To find the $y-$intercept, you can consider x equal to 0 and solve for y: $3x + 5y = -15 \rightarrow 3(0) + 5y = -15 \rightarrow 5y = -15 \rightarrow y = -3$. The $y-$intercept is -3.

Example 3. Find the $x-$ and $y-$intercepts of line $-8x + 16y = 64$.

Solution: To find the $x-$intercept, you can consider y equal to 0 and solve for x: $-8x + 16y = 64 \rightarrow -8x + 16(0) = 64 \rightarrow -8x = 64 \rightarrow x = -8$. The $x-$intercept is -8. To find the $y-$intercept, you can consider x equal to 0 and solve for y: $-8x + 16y = 64 \rightarrow -8(0) + 16y = 64 \rightarrow 16y = 64 \rightarrow y = 4$. The $y-$intercept is 4.

Graph an Equation in the Standard Form

- The linear equations' standard form is $Ax + By = C$. In the standard form of the equation, the letters a, b, and c are all substituted with real numbers. The letter x refers to the independent variable and the letter y refers to the dependent variable.
- If an equation is expressed in standard form, then you can't find the slope and y-intercept that you need for graphing at first glance at the equation. In such a case, you must use a method to find these values.
- There are 2 different methods to graph a line in standard form. The first way is to convert the equation to slope-intercept form ($y = mx + b$) and then graph it. The second way is to find x and y −intercepts of the line in standard form and connect 2 intercepts and draw the line.
- The easiest way to graph an equation's line in standard form is by identifying intercepts. Keep in mind that at the y −intercept, the coordinate of x is equal to zero and that at the x-intercept, the coordinate of y is equal to zero. When you want to find the y −intercept, put x equal to zero and solve for y. When you want to find the x −intercept, set y equal to zero and solve for x. Then find the 2 intercepts on the coordinate plane and draw the line on the graph.

Example:

Graph the following equation: $6x - 4y = 24$

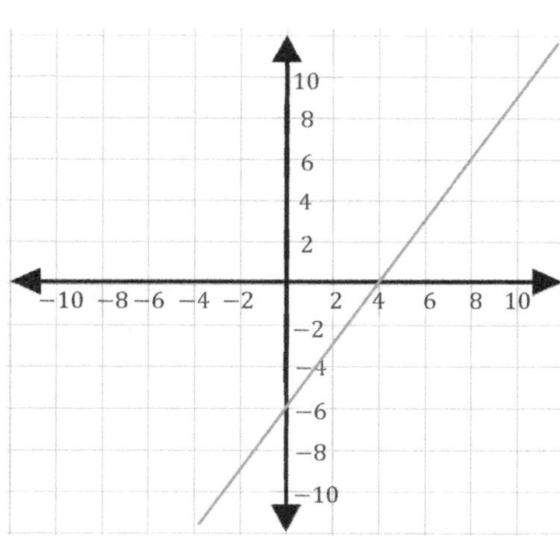

Solution: First, find the x −intercept. Consider $y = 0$ and solve for x: $6x - 4y = 24 \rightarrow 6x - 4(0) = 24 \rightarrow 6x = 24 \rightarrow x = 4$. The x −intercept is 4 and its coordinates are $(4,0)$. Now, find the y −intercept. Consider $x = 0$ and solve for y: $6x - 4y = 24 \rightarrow 6(0) - 4y = 24 \rightarrow -4y = 24 \rightarrow y = -6$. The y −intercept is -6 and its coordinates are $(0,-6)$. Now, find $(4,0)$ and $(0,-6)$ on the coordinate plane, and draw the line between these two points on the graph.

bit.ly/41Luqdp

Equations of Horizontal and Vertical lines

- Horizontal lines have a slope of 0. Thus, in the slope-intercept equations $y = mx + b$, $m = 0$, the equation becomes $y = b$, where b is the y −coordinate of the y −intercept.
- Similarly, in the graph of a vertical line, x only takes one value. Thus, the equation for a vertical line is $x = a$, where a is the value that x takes.

Examples:

Example 1. Write an equation for the horizontal line that passes through (6,2).

Solution: Since the line is horizontal, the equation of the line is in the form of:

$y = b$.

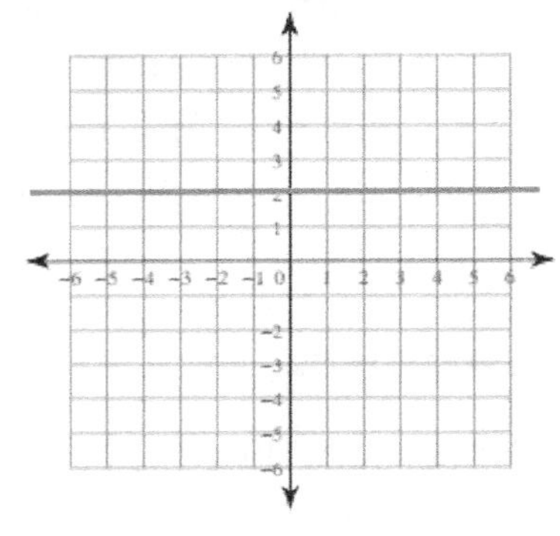

Where y always takes the same value of 2.

Thus, the equation of the line is:

$y = 2$.

Example 2. Write an equation for the vertical line that passes through (−3,5).

Solution: Since the line is vertical, the equation of the line is in the form of:

$x = a$.

Where x always takes the same value of −3.

Thus, the equation of the line is:

$x = -3$.

Graph a Horizontal or Vertical line

- When you graph a line, you usually need a point and the slope of the line to draw the line on the coordinate plane. But there are also exceptions that are called horizontal and vertical lines.
- The horizontal line is a kind of straight line that extends from left to right. An example of a horizontal line can be the x −axis. The horizontal line's slope is always zero because slope can be defined as "rise over run", and a horizontal line's rise is 0. Since the answer of dividing zero by any number is equal to zero, a horizontal line's slope always is zero. Horizontal lines are always parallel to the x −axis and they are written in the form $y = a$ where a is a real number.
- The vertical line is a kind of straight line that extends up and down. An example of a vertical line can be the y −axis. The vertical line's slope is always undefined because the run of a vertical line is zero. In fact, a number's quotient divided by zero is undefined, so a vertical line's slope is always undefined. Vertical lines are always parallel to the y −axis and they are written in the form $x = a$ where a is a real number.

Examples:

Example 1. Graph this equation: $y = -5$

Solution: $y = -5$ is a horizontal line and this equation tells you that every y −value is -5. You can consider some points that have a y −value of -5, then draw a line that connects the points.

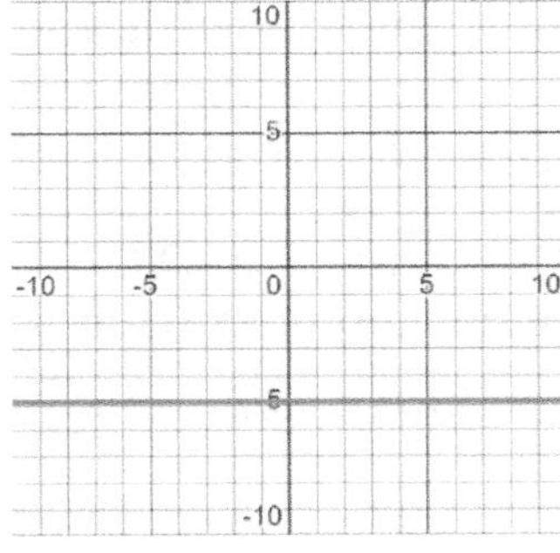

Example 2. Graph this equation: $x = 3$

Solution: $x = 3$ is a vertical line and this equation tells you that every x −value is 3. You can consider some points that have an x −value of 3, then draw a line that connects the points.

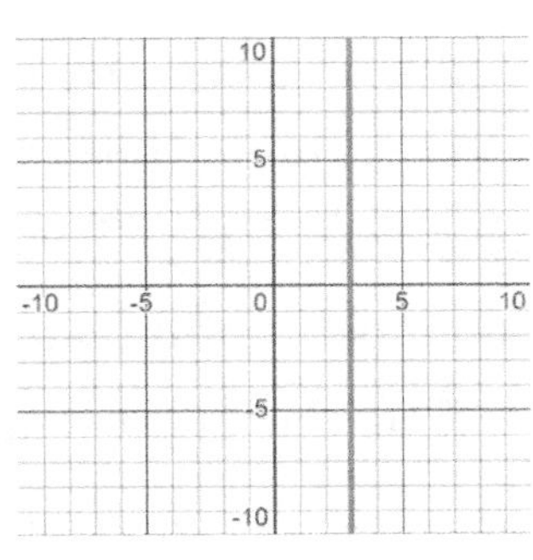

Graph an Equation in Point-Slope Form

- The point-slope form uses a slope and a point on a straight line to represent this line's equation. In fact, a line's equation with slope m that passes through the point (x_1, y_1) can be found by the point-slope form.
- The point-slope form's equation is $y - y_1 = m(x - x_1)$. In this equation, (x_1, y_1) is a point on the straight line and m is the line's slope.
- To graph an equation in point-slope form, follow these steps:
    - 1st step: First, check the point-slope form equation and be sure the equation uses subtraction the same as the point-slope form formula. If one side of the equation doesn't use subtraction operation, you should rewrite it with the subtraction sign.
    - 2nd step: Find a random point on the straight line and the slope of the line. The slope formula is $m = \frac{change\ in\ y}{change\ in\ x}$.
    - 3rd step: Use the random point and the value of the slope to graph the line.

Example:

Graph the following line: $y - 3 = -\frac{1}{3}(x + 5)$

Solution: First, check the point-slope form equation. If one side of the equation doesn't use subtraction operation, you should rewrite it with the subtraction sign: $y - 3 = -\frac{1}{3}(x + 5) \rightarrow y - 3 = -\frac{1}{3}(x - (-5))$. Find a random point on the straight line and the slope of the line. You can use the point that is used in the formula: $(x_1, y_1) \rightarrow (-5,3)$. The slope is $-\frac{1}{3}$. Use the random point and the value of slope to graph the line: find the point $(-5,3)$ on the coordinate plane. The slope is $-\frac{1}{3}$ and is the same as $\frac{-1}{3}$. So, move down 1 unit and right 3 units to find the other point on the straight line: $(-2,2)$. Connect these two points and graph the line.

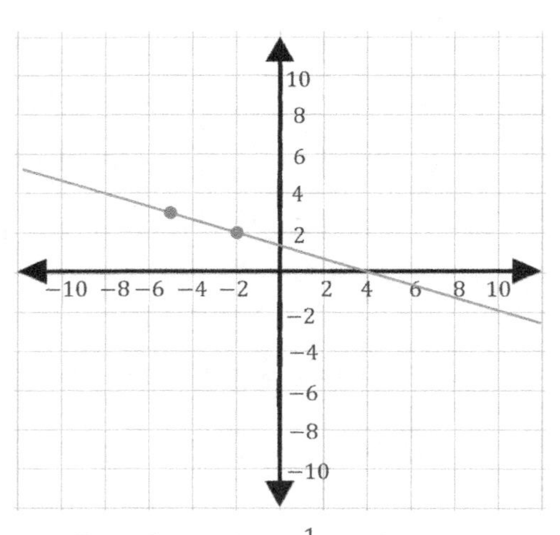

Equation of Parallel and Perpendicular Lines

- Parallel lines have the same slope.
- Perpendicular lines have opposite-reciprocal slopes. If the slope of a line is m, its Perpendicular line has a slope of $-\frac{1}{m}$.
- Two lines are Perpendicular only if the product of their slopes is negative 1. $m_1 \times m_2 = -1$.

Examples:

Find the equation of a line that is:

Example 1. Parallel to $y = 2x + 1$ and passes through the point (5,4).

Solution: The slope of $y = 2x + 1$ is 2. We can solve it using the "point-slope" equation of a line: $y - y_1 = 2(x - x_1)$ and then put in the point (5,4):
$y - 4 = 2(x - 5)$.
You can also write it in slope-intercept format: $y = mx + b$.
$y - 4 = 2x - 10 \rightarrow y = 2x - 6$.

Example 2. Perpendicular to $y = -4x + 10$ and passes through the point (7,2).

Solution: The slope of $y = -4x + 10$ is -4. The negative reciprocal of that slope is:$m = \frac{-1}{-4} = \frac{1}{4}$. So, the perpendicular line has a slope of $\frac{1}{4}$.
Then:$y - y_1 = (\frac{1}{4})(x - x_1)$ and now put in the point (7,2): $y - 2 = (\frac{1}{4})(x - 7)$.
Slope-intercept $y = mx + b$ form:$y - 2 = \frac{x}{4} - \frac{7}{4} \rightarrow y = \frac{1}{4}x + \frac{1}{4}$

Example 3. Parallel to $y = 5x - 3$ and passes through the point $(4, -1)$.

Solution: The slope of $y = 5x - 3$ is 5. We can solve it using the "point-slope" equation of a line:$y - y_1 = 5(x - x_1)$ and then put in the point $(4, -1)$: $y - (-1) = 5(x - 4)$.
You can also write it in slope-intercept format:$y = mx + b$. $y + 1 = 5x - 20 \rightarrow y = 5x - 21$

Compare Linear Function's Graph and Equations

- A linear function can be written in the form of $f(x) = mx + b$. In this form of equation m and b are real numbers.
- A linear graph provides a visual representation of a linear function and provides a straight line on the coordinate plane by connecting the points plotted on x and y coordinates.
- To compare a linear function's graph and linear equations, you should compare the slope or the y –intercept of them.
- To compare slopes of a linear function's graph and linear equations, find the change in y and the change in x between any 2 points on the graph's line. Then use the slope's equation ($m = \frac{change\ in\ y}{change\ in\ x}$) to find the value of m. After finding the value of m you can compare it with the value of m in a linear function.
- To compare the $y-$ intercept of the linear function's graph and linear equations, you should first find the value of b in the graph. To determine $y-$intercept (b), see at which point the line crosses the $y-$axis. Then compare the b –value of the graph with the b –value of the linear function.

Example:

Compare the slope of function A and function B.

Function A

Function B

$y = 2x - 6$

Solution: First, find the slope (m) of function A. Find the change in y and the change in x points (2,0) and (0, −6) on the line: $m = \frac{change\ in\ y}{change\ in\ x}: \frac{-6-0}{0-2} = 3$. Now, find the slope of function B. The equation of function B in the slope-intercept form is $y = 2x - 6$. Therefore, its slope is equal to 2. In the final step, compare the slopes. The slope of function A is 3 and is greater than the slope of function $B(2)$.

Graphing Absolute Value Equations

- The general form of an absolute value function (that is linear) is:

$$y = |mx + b| + c$$

- The vertex (the lowest or the highest point) is located at $(\frac{-b}{m}, c)$.
- A vertical line that divides the graph into two equal halves is: $x = -\frac{b}{m}$.
- To graph an absolute value equation, find the vertex and some other points by substituting some values for x and solving for y.

Example:

Graph $y = |x + 2|$.

Solution: Find the vertex $(\frac{-b}{m}, c)$.

According to the general form of an absolute value function:

$$y = |mx + b| + c.$$

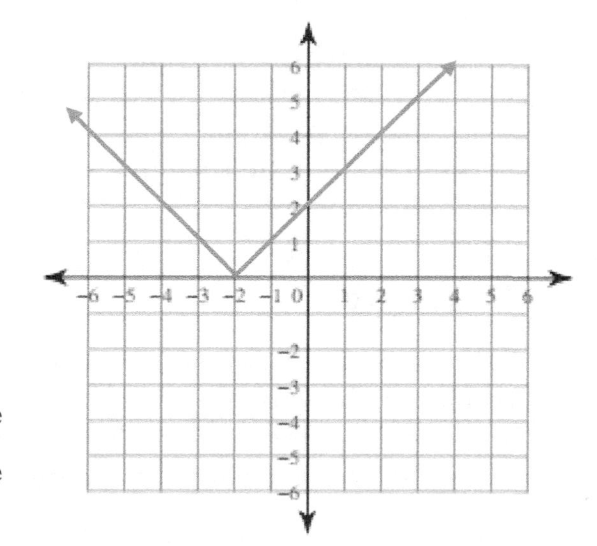

We have:

$$x = \frac{-b}{m} \rightarrow x = \frac{-2}{1} = -2.$$

And c is zero.

Then, the point $(-2,0)$ is the vertex of the graph and represents the center of the table of values. Create the table and plot the ordered pairs. Now, find the points and graph the equation.

Two-variable Linear Equation Word Problems

- A two-variable linear equation is a kind of linear equation that has two variables with the exponent 1 and the solutions of can two-variable linear equation can be expressed in ordered pairs like (x, y).
- A two-variable linear equation can be seen in various forms like standard form, point-slope form, and slope-intercept form:
  - The standard form of a two-variable linear equation is $ax + by = c$. In this form of the equation, a, b, and c are real numbers and x and y are the variables.
  - The slope-intercept form of a two-variable linear equation is as $y = mx + b$. In this form of the equation, m is the line's slope and b is the y −intercept of the line.
  - The point-slope form of a two-variable linear equation is as $y - y_1 = m(x - x_1)$. In this form of the equation, (x_1, y_1) is a given point on the line and m is the line's slope.
- To write and solve a two-variable linear equation word problem you can follow these steps:
  - 1st step: Read the question carefully and ask yourself what is given information and what is needed in the question.
  - 2nd step: Determine the unknowns in the question and show them by x, and y variables.
  - 3rd step: Change the words of the problem to the mathematical language or math expression.
  - 4th step: Make a two-variable linear equation using the given information in the problem.
  - 5th step: In the last step, solve the equation to find the value of the unknowns.

Example:

John is going to buy a box of candies as a gift for his friend. He can add as many candies as he wants to this box at a price of $3 per number. He also plans to buy a beautiful painting for $20 as a gift. Show in an equation how the total cost, y, depends on the number of candies, x.

Solution: The cost of each candy = $3 and the cost of the painting = $20. Here, the number of candies is unknown, and we consider the variable x for it, and we show the total cost with the variable y. Therefore, our two-variable linear equation will be as follows:

$y = 3x + 20$.

Chapter 4: Practices

Find the slope of each line.

1) $y = x - 5$
2) $y = 2x + 6$
3) $y = -5x - 8$
4) Line through $(2,6)$ and $(5,0)$
5) Line through $(8,0)$ and $(-4,3)$
6) Line through $(-2,-4)$ and $(-4,8)$

Solve.

7) What is the equation of a line with slope 4 and intercept 16 ?

____

8) What is the equation of a line with slope 3 and passes through point $(1,5)$?

____

9) What is the equation of a line with slope -5 and passes through point $(-2,7)$? _______________

10) The slope of a line is -4 and it passes through point $(-6,2)$. What is the equation of the line? _______________

11) The slope of a line is -3 and it passes through point $(-3,-6)$. What is the equation of the line? _______________

Sketch the graph of each linear inequality.

12) $y > 4x + 2$

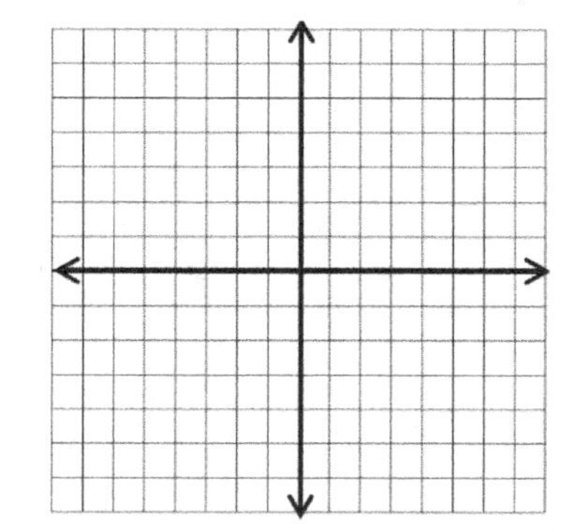

13) $y < -2x + 5$

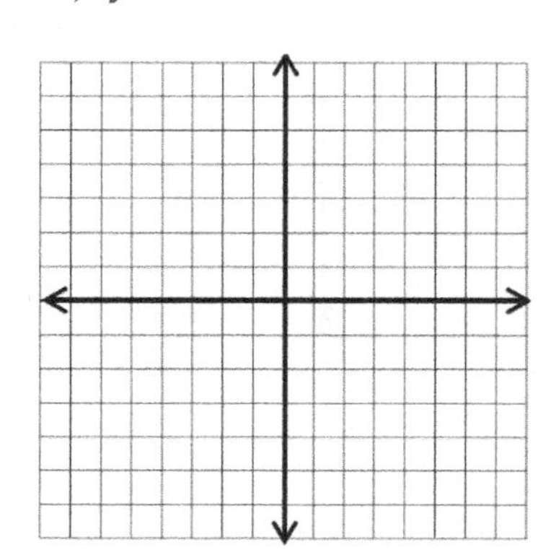

Write an equation of each of the following line in slope-intercept from.

14) ____________ 15) __________

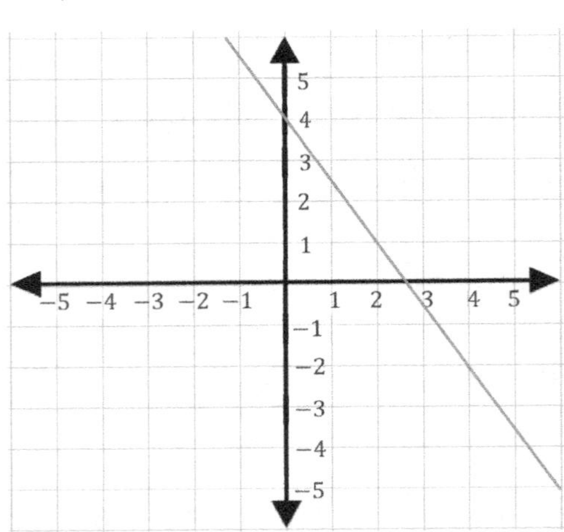

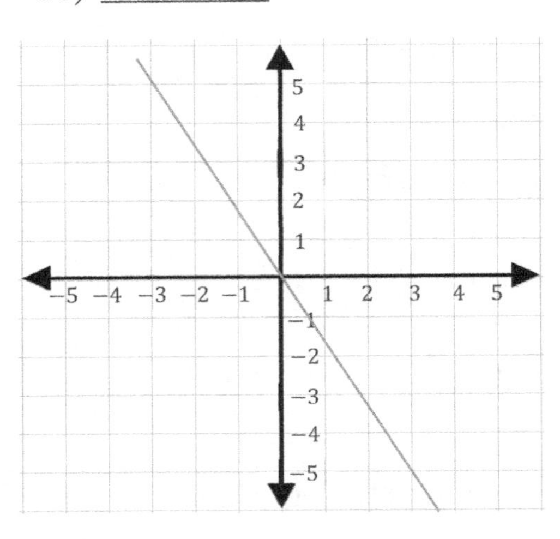

Find the equation of each line.

16) Through: $(6, -6)$, slope $= -2$
Point-slope form: ________
Slope-intercept form: ________

17) Through: $(-7,7)$, slope $= 4$
Point-slope form: _________
Slope-intercept form: ________

Write equation of the line in point-slope form.

18)

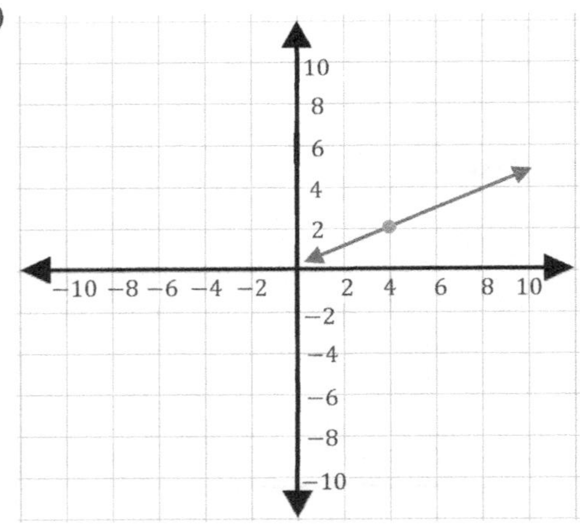

19)

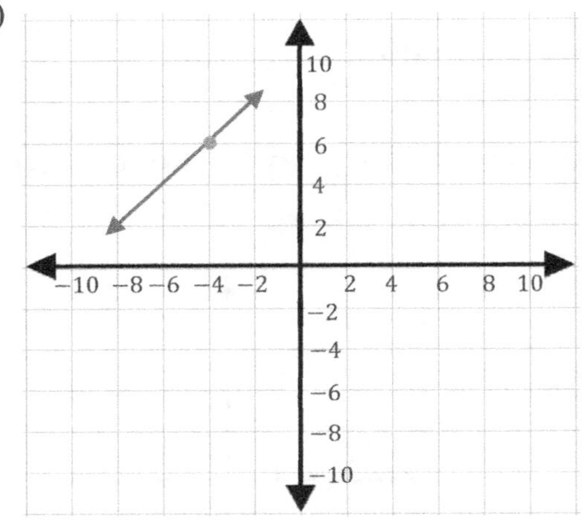

 **Find the x – intercept of each line.**

20) $21x - 3y = -18$
21) $20x + 20y = -10$
22) $8x + 6y = 16$
23) $2x - 4y = -12$

Graph each equation.

24) $4x - 5y = 40$

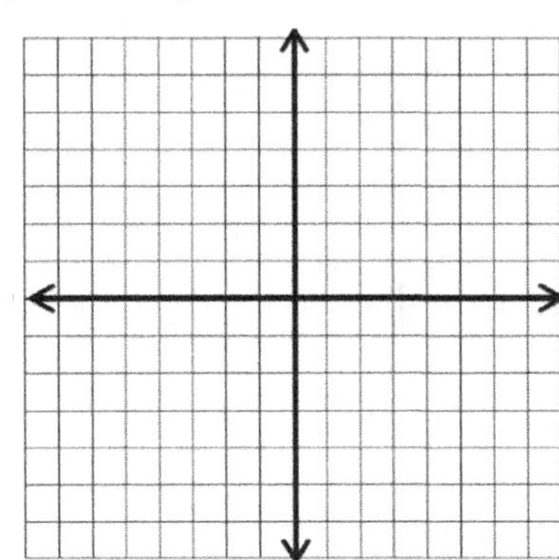

25) $9x - 8y = -72$

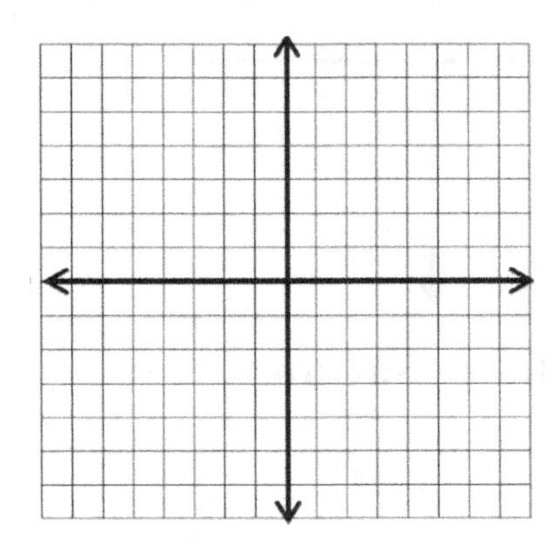

Find the equation of the following lines.

26) Write an equation for the horizontal line that passes through $(3, -5)$.
27) Write an equation for the horizontal line that passes through $(-4, 7)$.
28) Write an equation for the vertical line that passes through $(4, 0)$.
29) Write an equation for the vertical line that passes through $(0, -7)$.

Sketch the graph of each line.

30) Vertical line that passes through $(2,6)$.

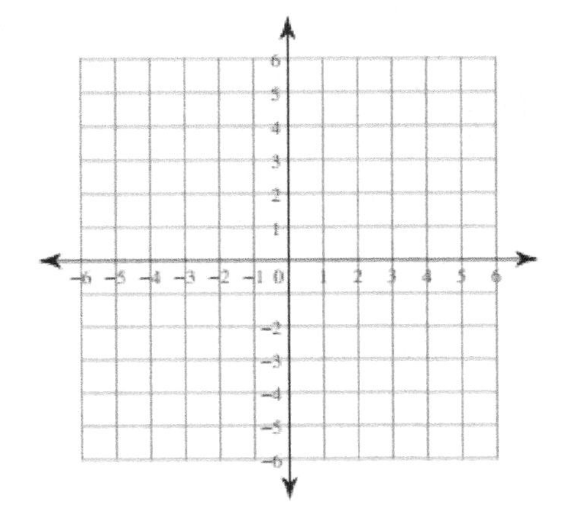

31) Horizontal line that passes through $(5,3)$.

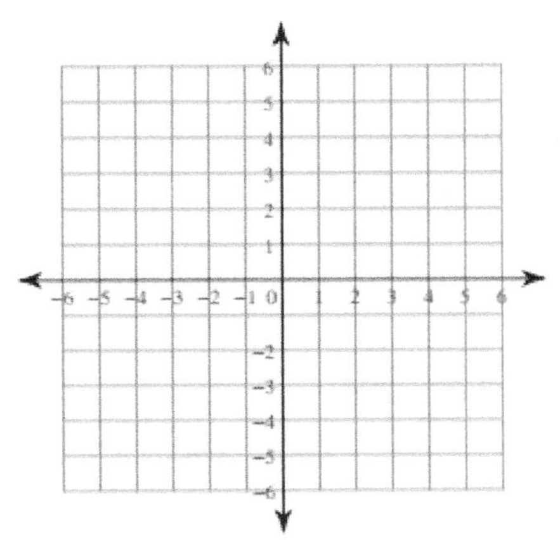

Effortless Math Education

Graph each equation.

32) $y + 3 = -\frac{1}{2}(x - 8)$

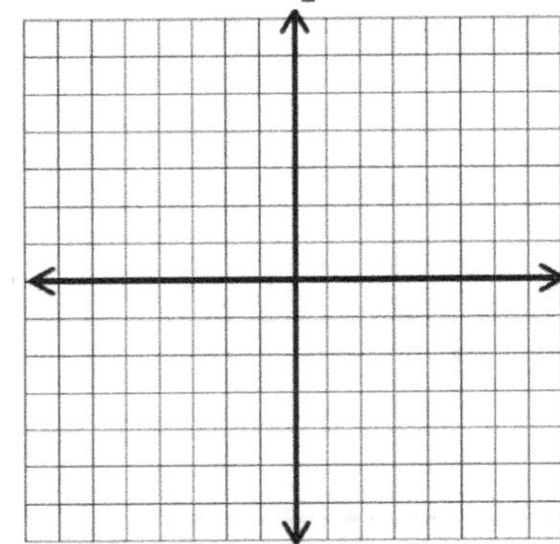

33) $y - 8 = -2(x - 1)$

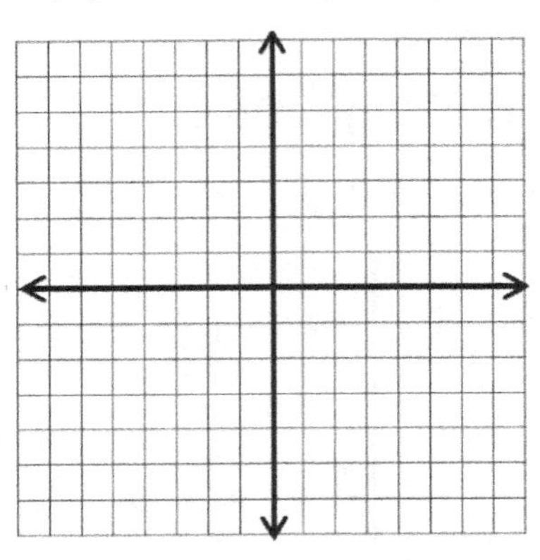

Find the equation of each line with the given information.

34) Through: $(4, 4)$

Parallel to $y = -6x + 5$

Equation: ________

35) Through: $(7,1)$

Perp. to $y = -\frac{1}{2}x - 4$

Equation: ________

36) Through: $(2,0)$

Parallel to $y = x$

Equation: ________

37) Through: $(0, -4)$

Perp. to $y = 2x + 3$

Equation: ________

38) Through: $(-1,1)$

Parallel to $y = 2$

Equation: ________

39) Through: $(3,4)$

Perp. to $y = -x$

Equation: ________

Compare the slope of the function A and function B.

40) Function A:

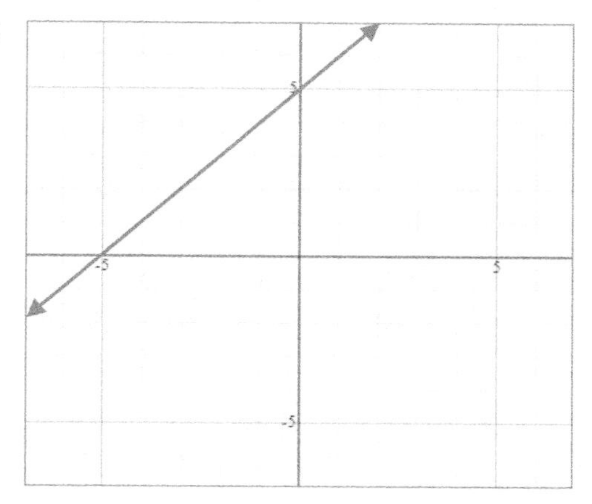

Function B: $y = 6x - 3$

41) Function A:

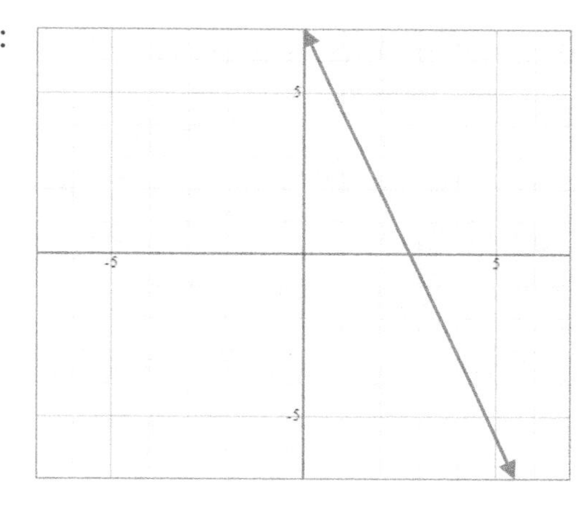

Function B: $y = -2.5x - 1$

42) Function A:

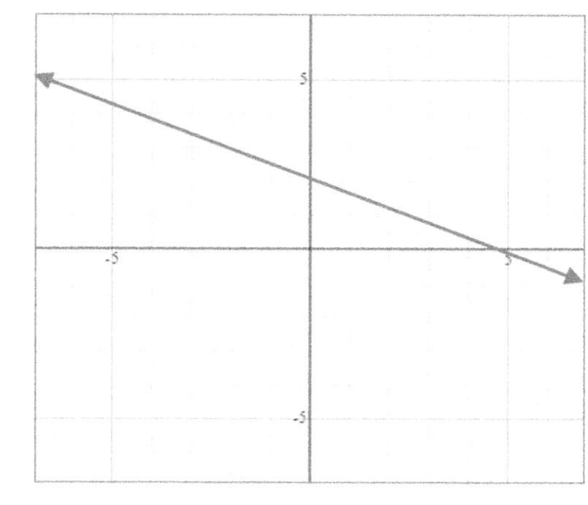

Function B: $y = 2x - 1$

Graph each equation.

43) $y = -|x| - 1$

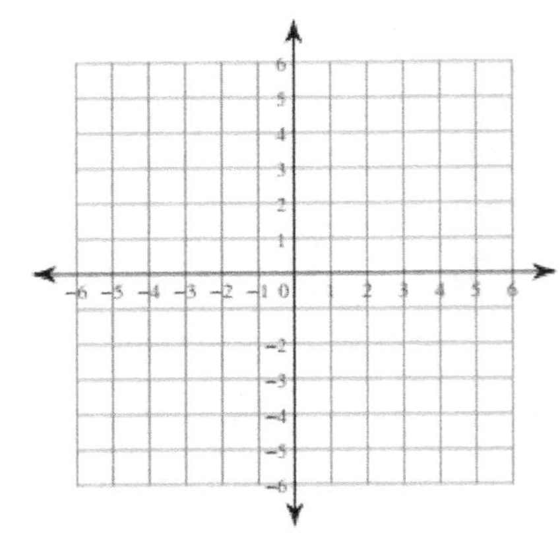

44) $y = -|x - 3|$

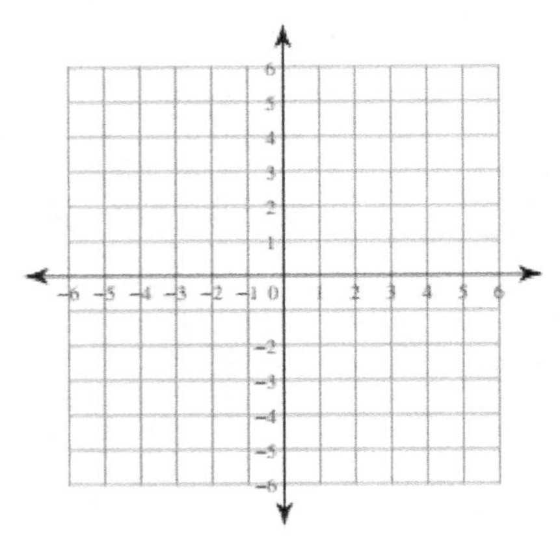

Solve.

45) John has an automated hummingbird feeder. He fills it to capacity, 8 fluid ounces. It releases 1 fluid ounce of nectar every day. Write an equation that shows how the number of fluid ounces of nectar left, y, depends on the number of days John has filled it, x.

46) The entrance fee to Park City is \$9. Additionally, skate rentals cost \$4 per hour. Write an equation that shows how the total cost, y, depends on the length of the rental in hours, x.

Chapter 4: Answers

1) 1
2) 2
3) -5
4) -2
5) $-\frac{1}{4}$
6) -6
7) $y = 4x + 16$
8) $y = 3x + 2$
9) $y = -5x - 3$
10) $y = -4x - 22$
11) $y = -3x - 15$

12) $y > 4x + 2$

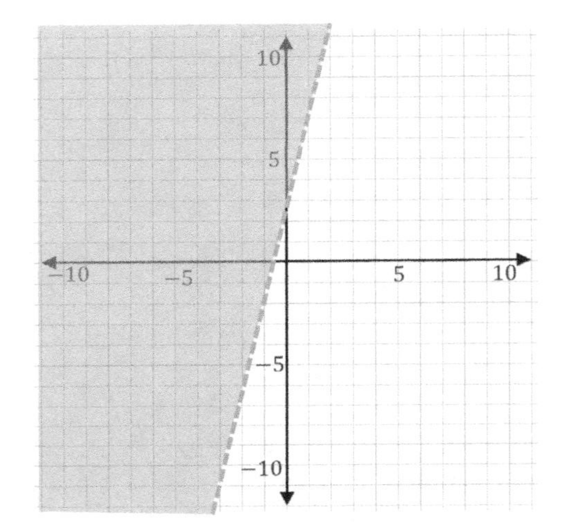

13) $y < -2x + 5$

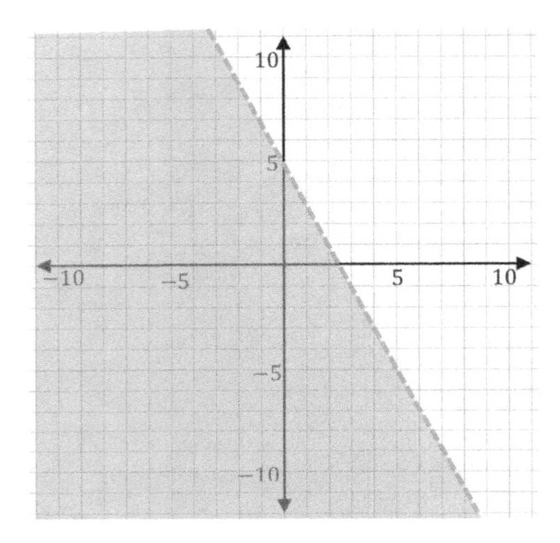

14) $y = -\frac{3}{2}x + 4$

15) $y = -\frac{5}{3}x$

16) Point-slope form: $y + 6 = -2(x - 6)$

Slope-intercept form: $y = -2x + 6$

17) Point-slope form: $y - 7 = 4(x + 7)$

Slope-intercept form: $y = 4x + 35$

18) $(y - 2) = \frac{1}{2}(x - 4)$
19) $(y - 6) = (x + 4)$
20) $-\frac{6}{7}$
21) $-\frac{1}{2}$
22) 2
23) -6

24)

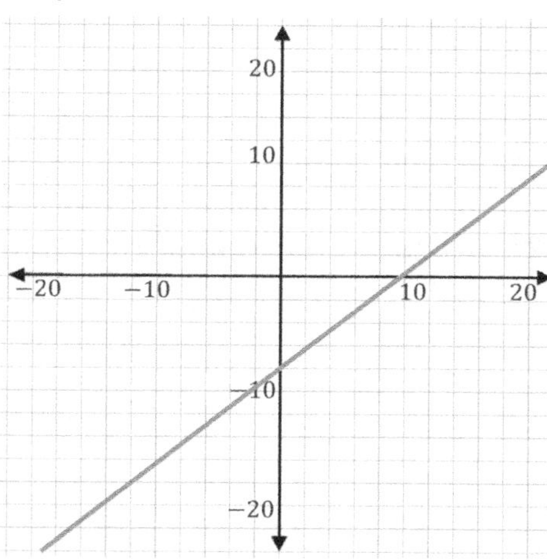

25)

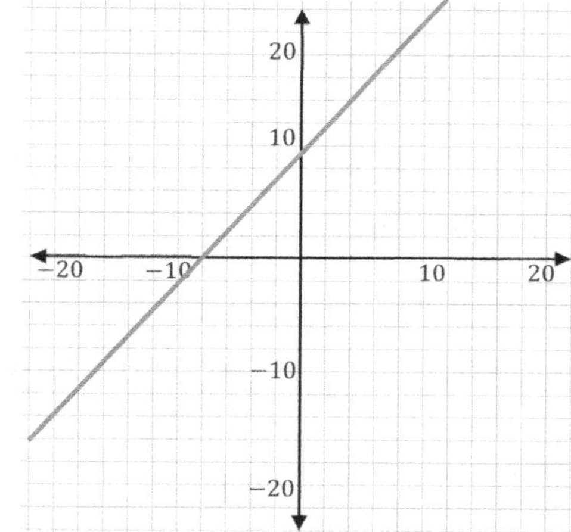

26) $y = -5$

27) $y = 7$

28) $x = 4$

29) $x = 0$

30)

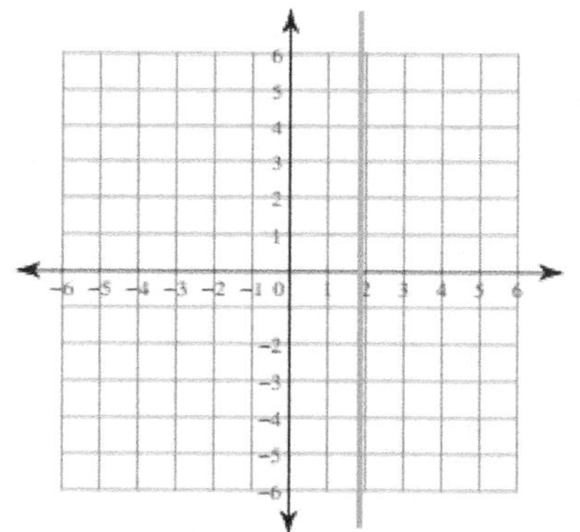

31)

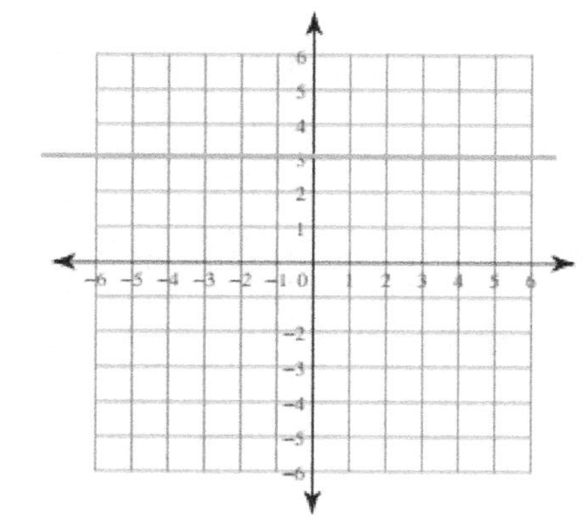

32)

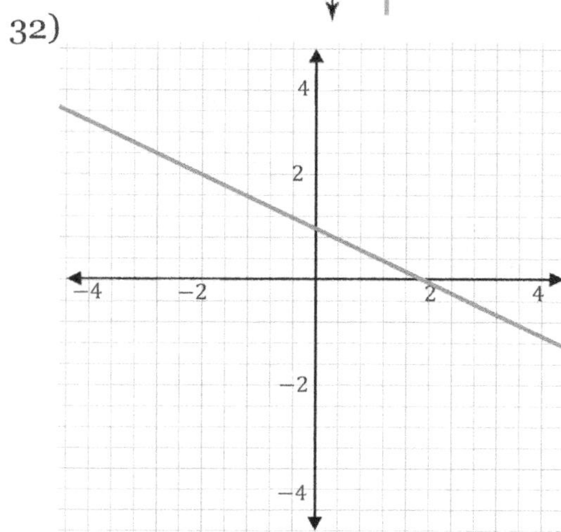

33)

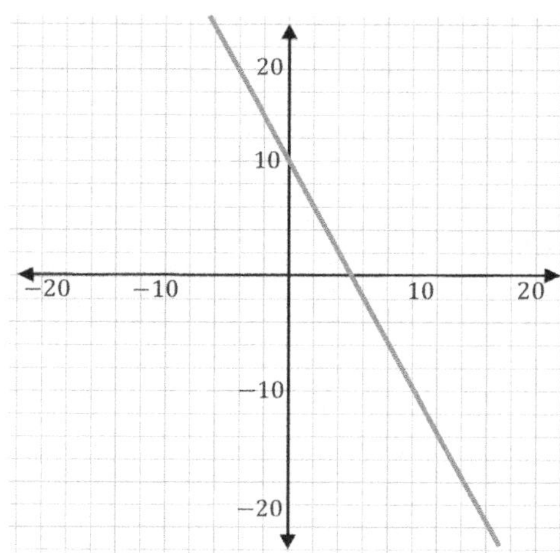

34) $y = -6x + 28$

35) $y = 2x - 13$

36) $y = x - 2$

37) $y = -\frac{1}{2}x - 4$

38) $y = 1$

39) $y = x + 1$

40) The slope of function A is 1 and is lower that than the slope of function B (6).

41) Two functions are parallel.

42) Two functions are perpendicular.

43) $y = -|x| - 1$

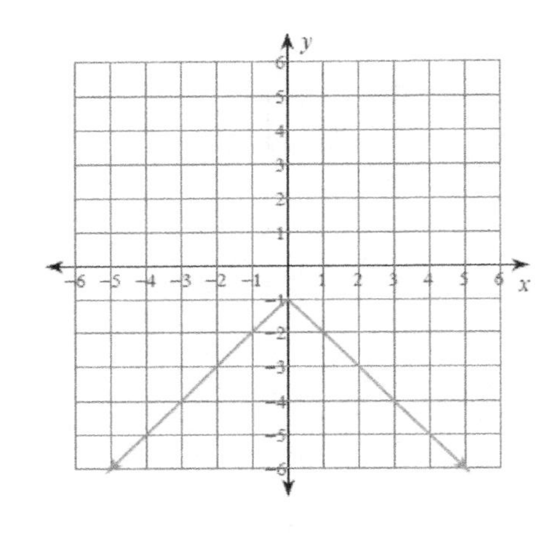

44) $y = -|x - 3|$

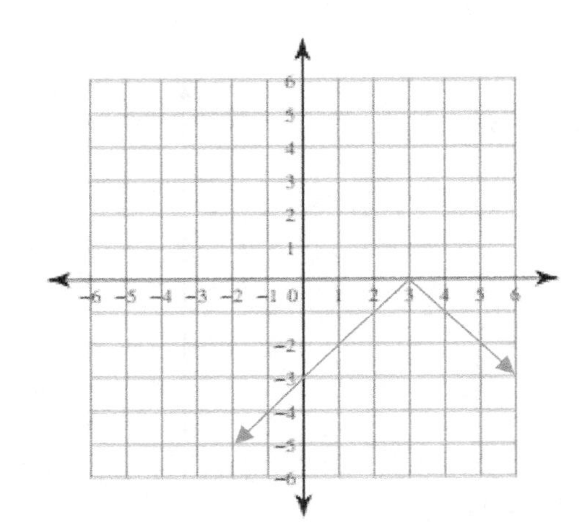

45) $y = -x + 8$

46) $y = 4x + 9$

CHAPTER

5 Inequalities and System of Equations

Math topics that you'll learn in this chapter:

- ☑ One–Step Inequalities
- ☑ Multi–Step Inequalities
- ☑ Compound Inequalities
- ☑ Write a Linear Inequality from a Graph
- ☑ Graph Solutions to Linear Inequalities
- ☑ Solve Advanced Linear Inequalities in Two-Variables
- ☑ Graph Solutions to Advanced Linear Inequalities
- ☑ Absolute Value Inequalities
- ☑ Systems of Equations
- ☑ Find the Number of Solutions to a Linear Equation
- ☑ Write a System of Equations Given a Graph
- ☑ Systems of Equations Word Problems
- ☑ Solve Linear Equations Word Problems
- ☑ Systems of Linear Inequalities
- ☑ Write Two-variable Inequalities Word Problems

One–Step Inequalities

- An inequality compares two expressions using an inequality sign. Inequality signs are: "less than" $<$, "greater than" $>$, "less than or equal to" $\leq$, and "greater than or equal to" $\geq$.
- You only need to perform one Math operation to solve the one-step inequalities.
- To solve one-step inequalities, find the inverse (opposite) operation is being performed. For dividing or multiplying both sides by negative numbers, flip the direction of the inequality sign.
- You can use "interval notation" to write the answers to inequalities. Interval notation is a way of representing the solutions of an inequality. Brackets and parentheses are used to indicate if the endpoints are included or excluded from the solution set. For example, consider the inequality $x > 2$. The solution set for this inequality is all real numbers greater than 2, extending to infinity. The interval notation for this solution set would be $(2, \infty)$, meaning that x is any number greater than 2 but does not include 2.
- If the inequality is $x \geq 2$, the solution set would include 2, and the interval notation for this solution set would be $[2, \infty)$, meaning that x is any number greater than or equal to 2, including 2. Keep in mind that we also have negative infinity that is represented by the symbol " $-\infty$".

Examples:

Example 1. Solve this inequality for x. $x + 5 \geq 4$

Solution: The inverse (opposite) operation of addition is subtraction. In this inequality, 5 is added to x. To isolate x we need to subtract 5 from both sides of the inequality.
Then: $x + 5 \geq 4 \rightarrow x + 5 - 5 \geq 4 - 5 \rightarrow x \geq -1$.
The solution is: $x \geq -1$ or $[-1, \infty)$

Example 2. Solve the inequality. $x - 3 > -6$

Solution: 3 is subtracted from x. Add 3 to both sides.
$x - 3 > -6 \rightarrow x - 3 + 3 > -6 + 3 \rightarrow x > -3$ or $[-3, \infty)$

Multi–Step Inequalities

- To solve a multi-step inequality, combine "like" terms on one side.
- Bring variables to one side by adding or subtracting.
- Isolate the variable.
- Simplify using the inverse of addition or subtraction.
- Simplify further by using the inverse of multiplication or division.
- For dividing or multiplying both sides by negative numbers, flip the direction of the inequality sign.

Examples:

Example 1. Solve this inequality. $8x - 2 \leq 14$

Solution: In this inequality, 2 is subtracted from $8x$. The inverse of subtraction is addition. Add 2 to both sides of the inequality:

$$8x - 2 + 2 \leq 14 + 2 \rightarrow 8x \leq 16.$$

Now, divide both sides by 8. Then:

$$8x \leq 16 \rightarrow \frac{8x}{8} \leq \frac{16}{8} \rightarrow x \leq 2.$$

The solution of this inequality is $x \leq 2$ or $(-\infty, 2]$

Example 2. Solve this inequality. $3x + 9 < 12$

Solution: First, subtract 9 from both sides: $3x + 9 - 9 < 12 - 9$.

Then simplify:

$$3x + 9 - 9 < 12 - 9 \rightarrow 3x < 3.$$

Now divide both sides by 3: $\frac{3x}{3} < \frac{3}{3} \rightarrow x < 1$ or $(-\infty, 1)$

Example 3. Solve this inequality. $-5x + 3 \geq 8$

Solution: First, subtract 3 from both sides:

$$-5x + 3 - 3 \geq 8 - 3 \rightarrow -5x \geq 5.$$

Divide both sides by -5. Remember that you need to flip the direction of inequality sign.

$$-5x \geq 5 \rightarrow \frac{-5x}{-5} \leq \frac{5}{-5} \rightarrow x \leq -1 \text{ or } (-\infty, -1]$$

Compound Inequalities

- A compound inequality includes two or more inequalities that are separated by the words "and" or "or".
- To solve compound inequalities, isolate the variable, flip the sign for negative division, and simplify using inverse operations.
- To solve compound inequality with the word "and," you must look for numbers that are solutions for all inequalities or the intersection of the inequalities.
- To solve a compound inequality with the word "or", first solve each inequality. Then graph the solutions. To find the solution to the compound inequality, we look at the graphs of each inequality, find the numbers that belong to each graph and put all those numbers together.
- To combine the solution sets of two or more inequalities, use union symbol "∪". For example, consider the inequalities $x < -5$ and $x > 2$. The solution sets for these inequalities can be represented as $(-\infty, -5)$ and $(2, \infty)$ respectively in interval notation. The union of these two intervals represents the combined solution set $x < -5$ or $x > 2$, which is $(-\infty, -5) \cup (2, \infty)$.

Examples:

Example 1. Solve. $6 < 3x \leq 24$

Solution: To solve this inequality, divide all sides of the inequality by 3. This simplifies the inequality as follows: $2 < x \leq 8$ or $(2,8]$.

Example 2. Solve. $x - 5 < -9$ or $\frac{x}{5} > 3$

Solution: Solve each inequality by isolating the variable:
$x - 5 < -9 \rightarrow x - 5 + 5 < -9 + 5 \rightarrow x < -4.$
Then:
$\frac{x}{5} > 3 \rightarrow \frac{x}{5} \times 5 > 3 \times 5 \rightarrow x > 15.$

The solution to these two inequalities is:
$x < -4$ or $x > 15$ or $(-\infty, -4) \cup (15, \infty)$

Write a Linear Inequality from a Graph

- When 2 linear expressions are compared by the inequality symbols $(<, >, \leq,)$ they make linear inequalities.
- The graph of inequality is represented by a dashed line or solid line, and one shaded side. With clues from a linear inequality graph and your information about linear relationships, you can find the equation of linear inequality.
- To write a linear inequality from a graph, follow these steps:
  - 1st step: Look at the graph and determine whether the inequality line is a dashed line or a solid line. If it's a dashed line, the sign of inequality is $<$ or $>$. If it's a solid line, the sign of inequality is $\leq$ or $\geq$.
  - 2nd step: Consider 2 points on the inequality line. Using these two points you can determine the equation of the inequality.
  - 3rd step: Find the inequality line's slope by these two points. To find slope you can use the slope's formula: $m = \frac{(y_2-y_1)}{(x_2-x_1)}$
  - 4th step: Put the value of slope and a point into the formula $y = mx + b$. In this formula, m is the slope of the line and (x, y) is a point on the line and the value of b is the y-intercept.
  - 5th step: Look at the graph's shaded part and determine whether y is less than the obtained equation or greater than the obtained equation. You can put a point from the shaded part into the equation to find the sign of inequality.

Example:

Write the slope-intercept form equation of the following graph.

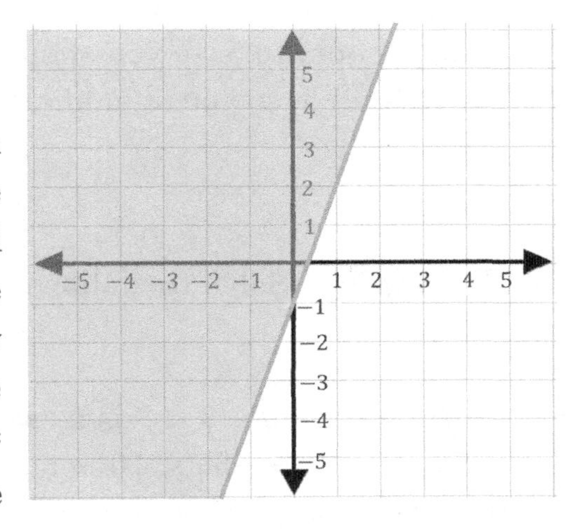

Solution: To write the inequality equation in slope-intercept form, you should find the y −intercept (b), and the slope (m), of the solid line. The value of b is -1 because the solid line passes through the y −axis at $(0, -1)$. Now consider 2 points on the solid line to find the slope. You can use $(0, -1)$ and $(-1, -4)$: $m = \frac{(y_2-y_1)}{(x_2-x_1)} = \frac{-4+1}{-1-0} = \frac{-3}{-1} = 3 \rightarrow m = 3$. Now use the value of b and m and put them into the slope-intercept form formula: $y = mx + b \rightarrow y = 3x - 1$. Determine the symbol of inequality: you have a solid line, and the shaded part is above the line. So, the equation of the inequality is as follows: $y \geq 3x - 1$.

Graph Solutions to Linear Inequalities

- A math statement that compares 2 expressions by an inequality sign is called an inequality. In inequalities, an inequality's expression can be greater or less than the other expression. The special sign should be used in inequalities $(\leq, \geq, >, <)$.
- Inequalities that you can solve using only one step are called one-step inequalities. Inequalities that you should take two steps to solve are called two-step inequalities.
- To solve one-step and two-step linear inequalities for a variable, you should use inverse operations to undo the operations and isolate the variable in the inequality. Remember you should do the same operation for two sides of the inequality. Reverse the inequality symbol's direction when you multiply or divide by a negative number.
- To graph an inequality, include a number using a filled-in circle and exclude a number using an open circle.

Examples:

Example 1. Solve the following inequality and graph the solution.

$$m + 3 \geq 8$$

Solution: Solve for m: $m + 3 \geq 8 \rightarrow m \geq 8 - 3 \rightarrow m \geq 5$. Now graph $m \geq 5$. The inequality $m \geq 5$ means that m can be any number more than or equal to 5. m can be equal to 5, so you should use a filled-in circle located on 5. Also, m can be more than 5, so you should also draw an arrow pointing to the right:

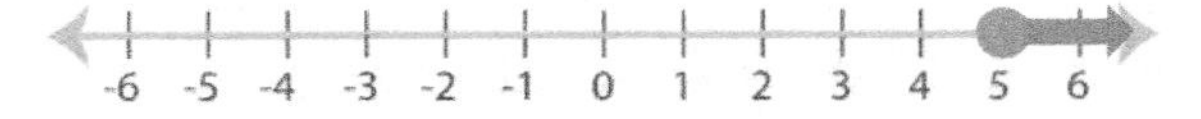

Example 2. Solve the following inequality and graph the solution.

$$-5q - 3 < 7$$

Solution: $-5q - 3 < 7$ is a two-step inequality. First, solve for q: $-5q - 3 < 7 \rightarrow -5q < 3 + 7 \rightarrow -5q < 10 \rightarrow q > -2$. Now graph $q > -2$. The inequality $q > -2$ means that q can be any number more than -2. q can't be equal to -2, so you should use an open circle located on -2. Also, q can be more than -2, so you should also draw an arrow pointing to the right:

Solve Advanced Linear Inequalities in Two-Variables

- The general form of a linear inequality in two-variable is as

  $$Ax + By < C,$$

  Where the sign less than ($<$) can be any of greater than ($>$), less than and equal to ($\leq$), greater than and equal to ($\geq$), or not equal to ($\neq$).
- To solve a linear inequality, convert the given inequality to the general form.
- Use the inverse operations to undo the inequality operations for both sides of the inequality to arrive at the general form.
- To solve a linear inequality in two-variable in general form all ordered pairs like (x, y) produce a true statement when the values of x and y are substituted into the inequality.

Examples:

Example 1. Solve the inequality $3x - 2 \geq 4y + x$.

Solution: First, convert to the general form. Add 2 to both sides of the inequality. So,

$$3x - 2 + 2 \geq 4y + x + 2 \rightarrow 3x \geq 4y + x + 2.$$

Subtract $4y$ from both sides as $3x - 4y \geq 4y + x + 2 - 4y \rightarrow 3x - 4y \geq x + 2$.

Also, subtract x from the sides:

$$3x - 4y \geq x + 2 \rightarrow 3x - 4y - x \geq x + 2 - x \rightarrow 2x - 4y \geq 2.$$

The answer to this inequality is all ordered pairs in the form of (x, y) where $y \leq \frac{x-1}{2}$. That is, $\left\{(x, y) | x \in \mathbb{R}, y \leq \frac{x-1}{2}\right\}$.

Example 2. Solve the inequality $y + 3 < x + 1$.

Solution: Convert to the general form. Subtract 1 from both sides. So,

$$y + 3 - 1 \geq x + 1 - 1 \rightarrow y + 2 < x.$$

Now, subtract y from both sides $y + 2 < x \rightarrow x - y > 2$. The answer to this inequality is all ordered pairs in the form of (x, y) where $x > y + 2$. That is, $\{(x, y) | y \in R, x > y + 2\}$.

Graph Solutions to Advanced Linear Inequalities

- Use the inverse operations to undo the inequality operations for both sides of the inequality.
- For dividing or multiplying both sides by negative numbers, flip the direction of the inequality sign.
- Continue simplifying until the variable is on one side of the inequality and the other components on the other side.
- To graph an inequality, use a filled-in circle to represent a number and an open circle to remove the number from the graph.

Examples:

Example 1. Solve the inequality $3x + 1 \geq 3 - x$.

Solution: Subtract 1 from both sides $3x + 1 - 1 \geq 3 - x - 1 \rightarrow 3x \geq 2 - x$. Now, add x to the sides of the obtained inequality. So, $3x + x \geq 2 - x + x \rightarrow 4x \geq 2$. Finally, divide by 4 and simplify. Therefore, $4x \geq 2 \rightarrow x \geq \frac{1}{2}$.

To graph, put a filled-in circle instead of the point $\frac{1}{2}$ on the real number axis and draw a line to positive infinity. As follow:

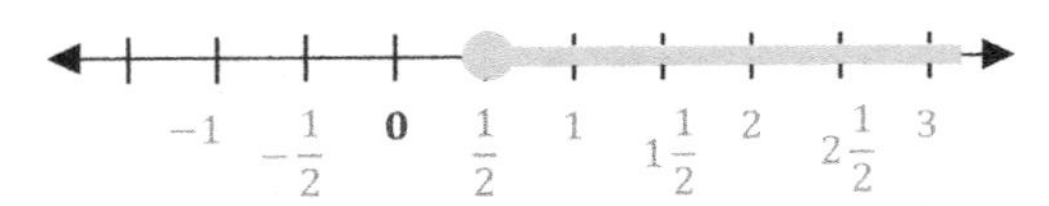

Example 2. Solve the inequality $\frac{x+3}{-3} < 2x + 1$.

Solution: Multiply -3 by both sides and since -3 is a negative number, then flip the direction of the inequality sign. So, $-3\left(\frac{x+3}{-3}\right) > -3(2x + 1) \rightarrow x + 3 > -6x - 3$. Add $6x$ to the sides of the inequalities: $x + 3 + 6x > -6x - 3 + 6x \rightarrow 7x + 3 > -3$. Now, subtract 3, $7x + 3 - 3 > -3 - 3 \rightarrow 7x > -6$. Finally, divide both sides of the inequality by 7. Therefore, $7x > -6 \rightarrow x > -\frac{6}{7}$. To graph, put an open circle instead of the point $-\frac{6}{7}$ on the axis and draw a line to $+\infty$. We have:

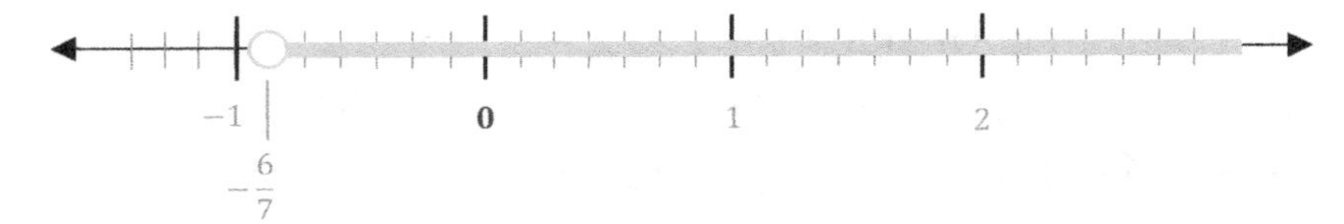

Absolute Value Inequalities

- An absolute value inequality includes an absolute value $|a|$ and a sign of inequality ($<, >, \leq, \geq$).
- To solve an absolute value inequality, change it from an absolute value to a simple inequality.
- The method of transforming absolute value inequality into simple inequality depends on the direction that the inequality refers to. Depending on the direction of the inequality, use one of the following methods:
  - ❖ To solve x in the inequality $|ax + b| < c$, you must solve $-c < ax + b < c$.
  - ❖ To solve x in the inequality $|ax + b| > c$ you must solve $ax + b > c$ and $ax + b < -c$.

Examples:

Example 1. Solve. $|2x - 1| \leq 5$.

Solution: Since the inequality sign is $\leq$, rewrite the inequality to:

$$-5 \leq 2x - 1 \leq 5.$$

Then, solve the inequality:

$$-5 \leq 2x - 1 \leq 5 \rightarrow -4 \leq 2x \leq 6$$

Now, divide each section by 2:

$$-4 \leq 2x \leq 6 \rightarrow \frac{-4}{2} \leq \frac{2x}{2} \leq \frac{6}{2} \rightarrow -2 \leq x \leq 3$$

You can also write this solution using the interval symbol: $[-2,3]$

Example 2. Solve. $|x + 3| > 11$.

Solution: Since the inequality sign is $>$, rewrite the inequality to:

$$x + 3 > 11 \text{ or } x + 3 < -11.$$

Now, simplify both inequalities:

$x + 3 > 11 \rightarrow x > 8$ or $x + 3 < -11 \rightarrow x < -14$ or $(-\infty, -14) \cup (8, +\infty)$

System of Equations

- A system of equations contains two equations and two variables. For example, consider the system of equations: $x - y = 1$, $x + y = 5$.
- The easiest way to solve a system of equations is by the elimination method. The elimination method uses the addition property of equality. You can add the same value to each side of an equation.
- For the first equation above, you can add $x + y$ to the left side and 5 to the right side of the first equation: $x - y + (x + y) = 1 + 5$. Now, if you simplify, you get: $x - y + (x + y) = 1 + 5 \rightarrow 2x = 6 \rightarrow x = 3$. Now, substitute 3 for the x in the first equation: $3 - y = 1$. By solving this equation, $y = 2$.

Examples:

Example 1. What is the value of $x + y$ in this system of equations? $\begin{cases} x + 2y = 6 \\ 2x - y = -8 \end{cases}$

Solution: Solving a System of Equations by Elimination: multiply the first equation by (-2), then add it to the second equation.

$$\frac{\begin{matrix} -2(x + 2y = 6) \\ 2x - y = -8 \end{matrix}}{} \rightarrow \begin{matrix} -2x - 4y = -12 \\ 2x - y = -8 \end{matrix} \rightarrow -5y = -20 \rightarrow y = 4$$

Plug in the value of y into one of the equations and solve for x.

$$x + 2(4) = 6 \rightarrow x + 8 = 6 \rightarrow x = 6 - 8 \rightarrow x = -2.$$

Thus,

$$x + y = -2 + 4 = 2.$$

Example 2. What is the value of $y - x$ in this system of equations? $\begin{cases} -x + 3y = 2 \\ 3x - y = -2 \end{cases}$

Solution: Solving a System of Equations by Elimination: multiply the second equation by (3), then add it to the first equation.

$$\begin{matrix} -x + 3y = 2 \\ 3(3x - y = -2) \end{matrix} \rightarrow \begin{matrix} -x + 3y = 2 \\ 9x - 3y = -6 \end{matrix} \rightarrow 8x = -4 \rightarrow x = -\frac{1}{2}.$$

Plug in the value of y into one of the equations and solve for y.

$$-\left(-\frac{1}{2}\right) + 3y = 2 \rightarrow \frac{1}{2} + 3y = 2 \rightarrow 3y = \frac{3}{2} \rightarrow y = \frac{1}{2}.$$

Thus, $y - x = \frac{1}{2} - \left(-\frac{1}{2}\right) = \frac{1}{2} + \frac{1}{2} = 1.$

Find the Number of Solutions to a Linear Equation

- The linear equation is a kind of equation with the highest degree of 1. In other words, in a linear equation, there is no variable with an exponent more than 1. A linear equation's graph always is in the form of a straight line, it's called a 'linear equation'.
- The linear equation has no solution if by solving a linear equation you get a false statement as an answer.
- The linear equation has one solution if by solving a linear equation you get a true statement for a single value for the variable.
- The equation has infinitely many solutions if by solving a linear equation you get a statement that is always true.

Examples:

Example 1. How many solutions does the following equation have?

$$7-4p=-4p$$

Solution: Solve for p: $7-4p=-4p \rightarrow 7=+4p-4p \rightarrow 7=0$. $7=0$ is a false statement. The linear equation has no solution because by solving the linear equation you get a false statement as an answer.

Example 2. How many solutions does the following equation have?

$$12h=3h+27$$

Solution: Solve for h: $12h=3h+27 \rightarrow 12h-3h=27 \rightarrow 9h=27 \rightarrow h=3$. $h=3$ is a true statement for a single value for the variable. So, the linear equation has one solution because by solving a linear equation you get a true statement for a single value for the variable.

Example 3. How many solutions does the following equation have?

$$8n-2n=6n$$

Solution: Simplify the left side of the equation
$8n-2n=6n \rightarrow 6n=6n$. $6n=6n$ is a statement that is always true. So, the equation has infinitely many solutions because by solving a linear equation you get a statement that is always true.

Write a System of Equations Given a Graph

- A system of equations consists of 2 or more equations and tries to find common solutions to the equations. In fact, a system of linear equations is an equation set in which the same set of variables can satisfy it."
- To write a system of equations given a graph, first, you should know each line in the graph shows a linear equation, and the two equations make a system of equations.
- Look at the first line. To write its equation, find the slope (m), and y −intercept (b). To find the slope, you can use any two points on the line and plug the, in the slope equation: $m = \frac{change\ in\ y}{change\ in\ x}$. To determine y −intercept (b), see at which point the line crosses the y −axis. Then use the value of slope and y −intercept to construct the line's equation in slope-intercept form. You can find the second line's equation in the same way.

Example:

Write a system of equations for the following graph.

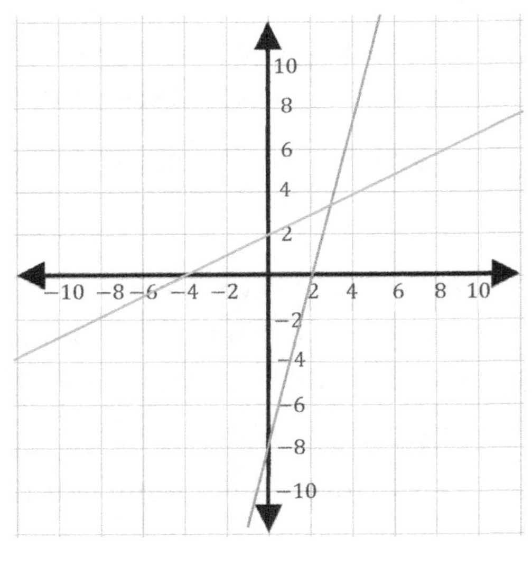

Solution: First, look at the green line. write its equation by identifying the slope (m) and y −intercept (b): to find the slope, you can use any two points on the line and plug in the slope equation: $m = \frac{change\ in\ y}{change\ in\ x}$.
Here choose the points $(0,2)$ and $(-4,0)$: $\frac{0-2}{-4-0} = \frac{1}{2}$.
The green line crosses the y −axis at $(0,2)$, so the y −intercept is $b = 2$. Now write the equation of the green line: $y = \frac{1}{2}x + 2$. You can find the red line's equation in the same way. Use the points $(0,-8)$ and $(2,0)$: $\frac{0+8}{2-0} = 4$. The red line crosses the y −axis at $(0,-8)$, so the y −intercept is $b = -8$. Now write the equation of the red line: $y = 4x - 8$. Therefore, the graph shows the following system of equations: $\begin{cases} y = \frac{1}{2}x + 2 \\ y = 4x - 8 \end{cases}$

Systems of Equations Word Problems

To solve systems of equations word problems:

- Find the key information in the problem that will help write two equations.
- Define two variables x and y.
- Write two equations using the variables.
- Choose a method (elimination, substitution, etc.) for solving the system of equations.
- Check your answers by substituting solutions into the original equations.
- Answer the questions in the real-world problems.

Example:

Tickets to a movie cost \$8 for adults and \$5 for students. A group of friends purchased 20 tickets for \$115. How many adult tickets did they buy?

Solution: Let x be the number of adult tickets and y be the number of student tickets. There are 20 tickets.

Then: $x + y = 20$. The cost of adults' tickets is \$8 and for students' it is \$5, and the total cost is \$115. So, $8x + 5y = 115$. Now, we have a system of equations:

$$\begin{cases} x + y = 20 \\ 8x + 5y = 115 \end{cases}.$$

To solve this system of equations, multiply the first equation by -5 and add it to the second equation: $-5(x + y = 20) = -5x - 5y = -100$.

$8x + 5y + (-5x - 5y) = 115 - 100 \rightarrow 3x = 15 \rightarrow x = 5 \rightarrow 5 + y = 20 \rightarrow y = 15$.

There are 5 adult tickets and 15 student tickets. Now, check your answers by substituting solutions into the original equations.

$x + y = 20 \rightarrow 5 + 15 = 20$, $8x + 5y = 115 \rightarrow 8(5) + 5(15) = 115 \rightarrow 40 + 75 = 115$.

The solutions are correct in both equations.

Solve Linear Equations' Word Problems

- A linear equation in one variable that you can solve in only one step is called a one-step equation.
- A linear equation in one variable that you should take two steps to solve is called a two-step equation.
- To solve word problems involving one-step and two-step linear equations, you can follow these steps:
  - 1st step: Read the whole problem carefully and determine what you are asked to find.
  - 2nd step: Look for keywords and put variables for the unknown amount.
  - 3rd step: Use the information you find and write an algebraic equation.
  - 4th step: Solve the equation. You should isolate the variable in your linear equation. Use simple math operations to isolate the variable. Remember when you use an operation for one side of the equation, you must also do the same for the other side. Once you've solved the linear equation, you've got the variable's value that makes the linear equation true.

Example:

Larry is in a chocolate shop and is going to buy some chocolates for his friends. He chooses 4 chocolates with a flower design and in addition, he also chooses some packs of three chocolates. If the total number of chocolates he has bought is 28, write an equation that you can use to find p, the number of packs of three chocolates. How many packs of three chocolates has Larry bought?

Solution: Larry chooses some packs of three chocolates and m is the number of packs of three chocolates. We can write it as follows: $3p$. He also chooses 4 chocolates with a flower design. The total number of chocolates he has bought is 28. Therefore, we can complete the equation as follows: $3p + 4 = 28$. The equation $3p + 4 = 28$ can be used to find how many packs of three chocolates Larry bought. Solve the equation for p: $3p + 4 = 28 \rightarrow 3p = 28 - 4 \rightarrow 3p \rightarrow 24 \rightarrow p = 24 \div 3 = 8 \rightarrow p = 8$. So, he bought 8 packs of three chocolates.

Systems of Linear Inequalities

- A system of linear inequalities is a set of two or more inequalities. Each of the inequalities is solved separately, and the common answer between these inequalities is the answer of the linear inequality system.
- If you cannot mentally find the answers to the inequalities, just draw on their axis and find the common ground.

Example:

Solve the following system of inequalities:

$$\begin{cases} 8x - 4y \leq 12 \\ 3x + 6y \leq 12 \\ y \geq 0 \end{cases}$$

Solution: As much as possible, we simplify each of the inequalities based on y:

$$\begin{cases} 8x - 4y \leq 12 \rightarrow -4y \leq -8x + 12 \rightarrow y \geq 2x - 3 \\ 3x + 6y \leq 12 \rightarrow 6y \leq -3x + 12 \rightarrow y \leq -\frac{x}{2} + 2 \rightarrow \\ y \geq 0 \end{cases} \begin{cases} y \geq 2x - 3 \\ y \leq -\frac{x}{2} + 2 \\ y \geq 0 \end{cases}$$

Draw a graph for each inequality. The answer to the system of inequalities is the common points between the graphs:

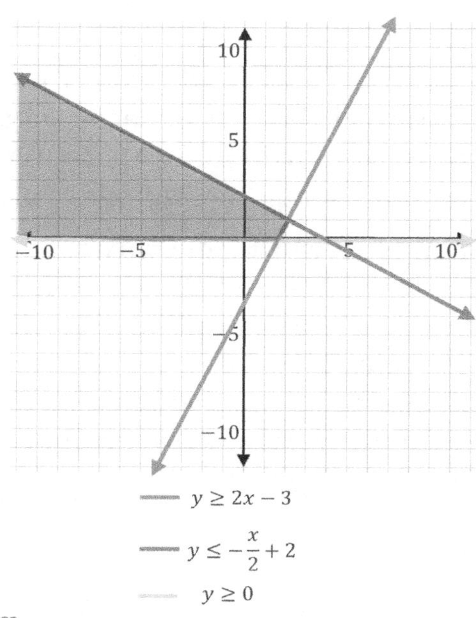

Write Two-variable Inequalities Word Problems

- The two-variable linear inequalities describe a not equal relationship between 2 algebraic statements that contain two different variables. A two-variables linear inequality is created when two variables are involved in the equation and inequality symbols ($<, >, \leq$, or $\geq$) are used to connect 2 algebraic expressions.
- The two-variables linear inequalities' solution can be expressed as an ordered pair (x, y). When the x and y values of the ordered pair are replaced in the inequality they make a correct expression.
- The method of solving a linear inequality is like the method of solving a linear equation. The difference between these two methods is in the symbol of inequality. You solve linear inequalities' word problems in the same way as linear equations' word problems:
  - 1st step: First, read the problem and try to write an inequality in words in a way that expresses the situation well.
  - 2nd step: Represent each of the given information with numbers, symbols, and variables.
  - 3rd step: Compare the inequality you've written in words with the given information that you've represented with numbers, symbols, and variables. Then rewrite what you've found out in the form of an inequality expression.
  - 4th step: Simplify both sides of the inequality. Once you find the values, you have one of these inequalities:
    - Strict inequalities: In this case, both sides of the inequalities can't be equal to each other.
    - Non-strict inequalities: In this case, both sides of the inequalities can be equal.

Example:

Sara plans to hold a party. She plans to order pizza and pasta from a restaurant for dinner. The cost of each pizza is \$45 and the cost of each pasta is \$30. She hopes to spend no more than \$400 on dinner. Write a linear inequality so that x represents the number of pizzas and y represents the number of pasta.

Solution: First, try to write an inequality in words in a way that expresses the situation well: The cost of the pizzas plus the cost of the pasta is no more than \$400. Now, represent each of the given information with numbers, symbols, and variables: The cost of the pizzas is \$45 times the number of pizzas, which is x. The product is $45x$. The cost of the pasta is \$30 times the number of pasta, which is y. The product is $30y$. No more than means less than or equal ($\leq$). Rewrite what you've found out in the form of an inequality expression: $45x + 30y \leq 400$.

Chapter 5: Practices

Solve each inequality and graph it.

1) $x - 2 \geq -2$

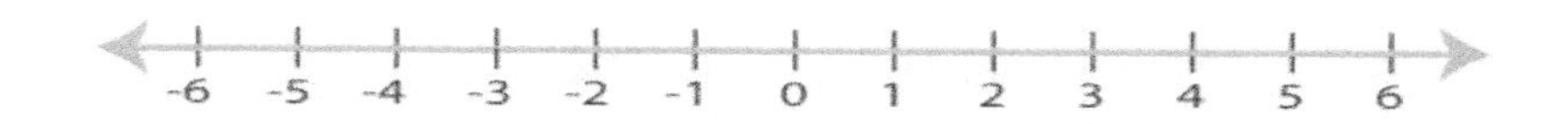

2) $2x - 3 < 9$

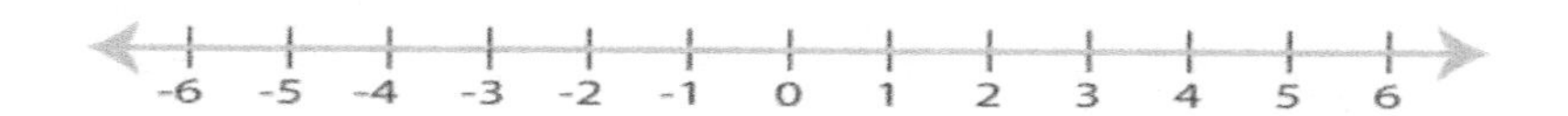

Solve each inequality.

3) $x + 13 > 4$

4) $x + 6 > 5$

5) $-12 + 2x \leq 26$

6) $-2 + 8x \leq 14$

7) $6 + 4x \leq 18$

8) $4(x + 3) \geq -12$

9) $2(6 + x) \geq -12$

10) $3(x - 5) < -6$

11) $10 + 5x < -15$

12) $6(6 + x) \geq -18$

13) $2(x - 5) \geq -14$

14) $6(x + 4) < -12$

15) $3(x - 8) \geq -48$

16) $-(6 - 4x) > -30$

17) $2(2 + 2x) > -60$

18) $-3(4 + 2x) > -24$

Solve each inequality.

19) $5x \leq 45$ and $x - 11 > -21$

20) $-7 < x - 9 < 8$

Write the slope-intercept from equation of the following graph.

21)

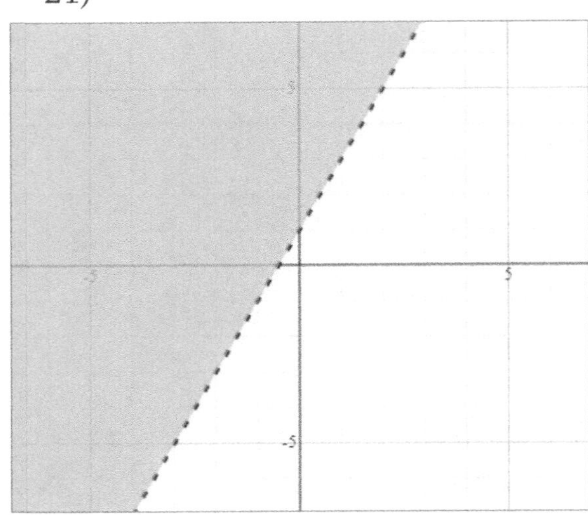

23)

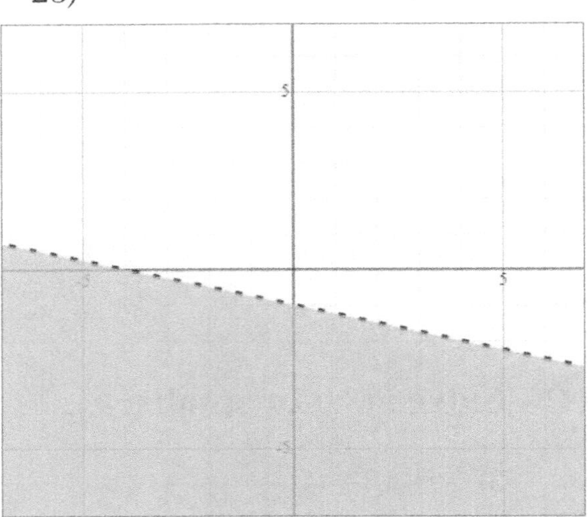

22)

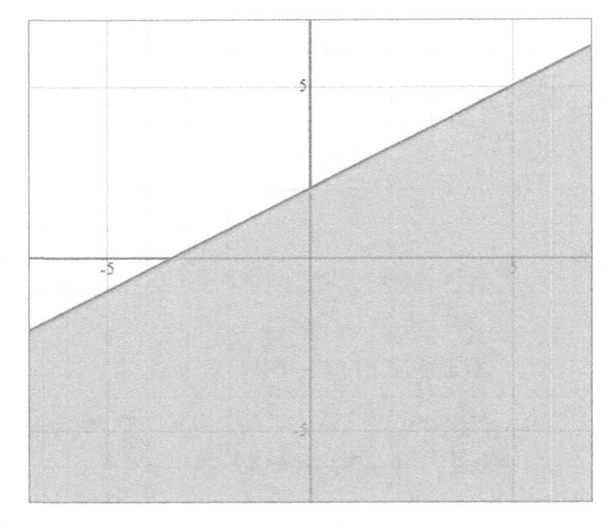

24)

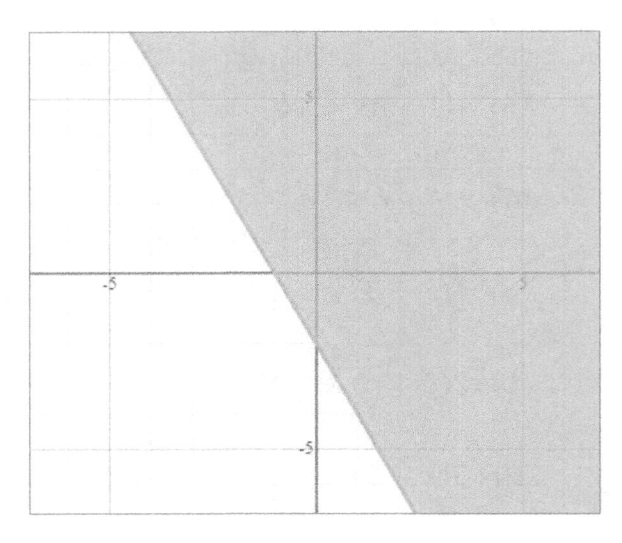

Solve the following inequality and graph the solution.

25) $10 + 6p \leq -2$

26) $-r + 8 \leq 4$

27) $-2f + 10 \geq 6$

28) $1 + 3p > 7$

Solve each inequality.

29) $8x - 3 \geq 4y + 2$

30) $4x - 3 \geq 5y + x$

31) $y \leq \frac{3}{2}x + 4$

32) $5x - 2y \leq 10$

Graph the solution of each inequality.

33) $7x + 3 \geq 1 - 2x$

34) $\frac{x+4}{-4} > 8x + 2$

Solve each inequality.

35) $|x| - 4 < 17$

36) $6 + |x - 8| > 15$

37) $\left|\frac{x}{2} + 3\right| > 6$

38) $\left|\frac{x+5}{4}\right| < 7$

Solve each system of equations.

39) $\begin{cases} -2x + 2y = -4 \\ 4x - 9y = 28 \end{cases}$ $x =$ $y =$

40) $\begin{cases} x + 8y = -5 \\ 2x + 6y = 0 \end{cases}$ $x =$ $y =$

41) $\begin{cases} 4x - 3y = -2 \\ x - y = 3 \end{cases}$ $x =$ $y =$

42) $\begin{cases} 2x + 9y = 17 \\ -3x + 8y = 39 \end{cases}$ $x =$ $y =$

How many solutions does the following equation have?

43) $4n = 8 + 5n$

44) $5 - 9f = -9f$

45) $0 = 3z - 3z$

46) $-9x + 2 = -9x$

47) $20 + 12y = 11y$

48) $10h - 2 = -4h$

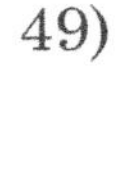

Write a system of equations for the following graph.

49)

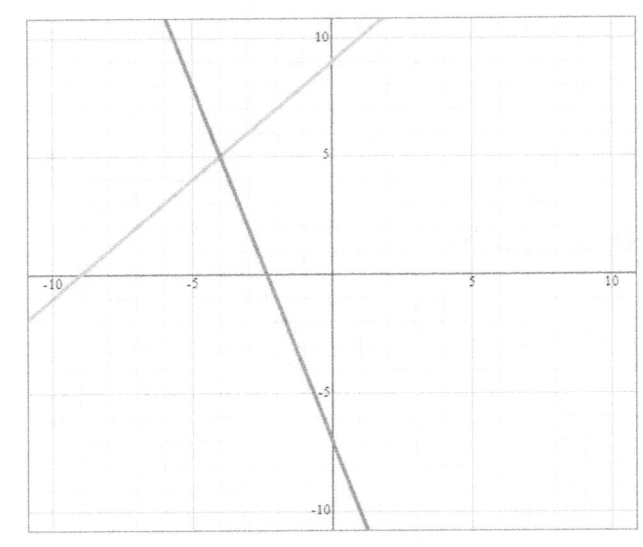

50)

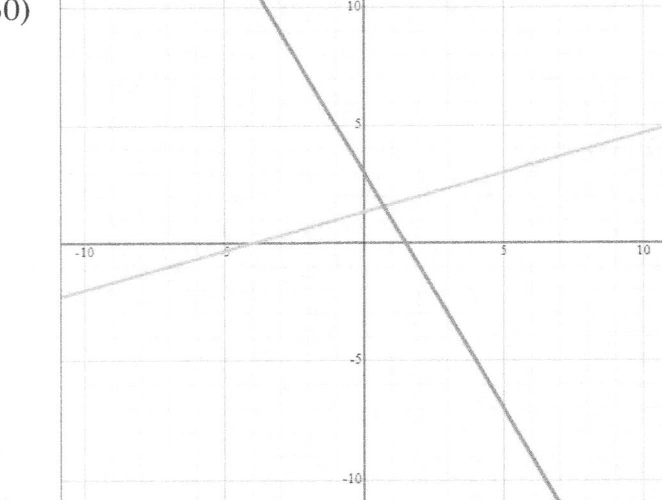

Solve each word problem.

51) The equations of two lines are $3x - y = 7$ and $2x + 3y = 1$. What is the value of x in the solution for this system of equations?

52) The perimeter of a rectangle is 100 feet. The rectangle's length is 10 feet less than 5 times its width. What are the length and width of the rectangle?

 **Solve each word problem.**

53) A golf club charges \$150 to join the club and \$15 for every hour using the driving range. Write an equation to express the cost C in terms of h hours playing tennis.

54) Susan is twice as old as Jane. In 4 years, Susan will be 24 years old. How old is Jane now?

55) A movie ticket costs \$7. Popcorn costs \$3 more than the ticket. If Alex bought 1 movie ticket and 1 popcorn, how much did he spend in total?

 **Solve each system of inequalities and graph them.**

56) $\begin{cases} x + 2y \le 3 \\ y - x \ge 0 \\ y \ge -2 \end{cases}$

57) $\begin{cases} x < 3 \\ x + y > -2 \\ y - 1 \le x \end{cases}$

Solve each word problem.

58) James used his first 2 tokens in Glimmer Arcade to play a Roll-and-Score game. Then he played his favorite game, Balloon Bouncer, over and over until he ran out of tickets. Balloon Bouncer costs 4 tokens per game and James started the game with a bucket of 38 tokens. Write an equation James can use to find how many games of Balloon Bouncer, g, he played.

59) Sara buys juice and soda for the party and wants to spend no more than \$46. The price of each bottle of soda is 3 dollars and each bottle of fruit juice is 1 dollar. Write the inequality in a standard form that describes this situation. Use the given numbers and variables below.

x = the number of bottles of soda

y = the number of bottles of juice

Effortless Math Education

Chapter 5: Answers

1) $x \geq 0$

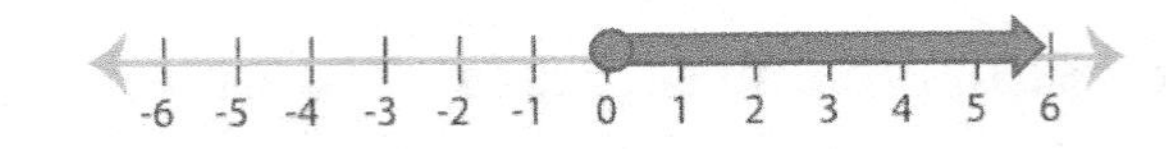

2) $x < 6$

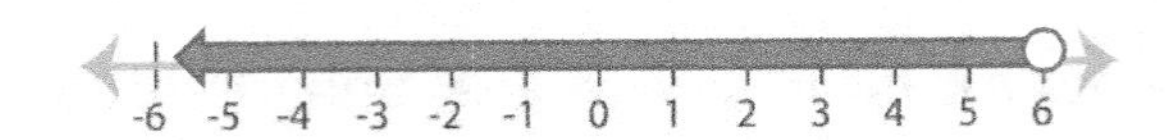

3) $x > -9$

4) $x > -1$

5) $x \leq 19$

6) $x \leq 2$

7) $x \leq 3$

8) $x \geq -6$

9) $x \geq -12$

10) $x < 3$

11) $x < -5$

12) $x \geq -9$

13) $x \geq -2$

14) $x < -6$

15) $x \geq -8$

16) $x > -6$

17) $x > -16$

18) $x < 2$

19) $-10 < x \leq 9$

20) $2 < x < 17$

21) $y > 2x + 1$

22) $y \leq \frac{3}{5}x + 2$

23) $y > -\frac{1}{4}x - 1$

24) $y \geq -2x - 2$

25) $p \leq -2$

26) $r \geq 4$

27) $f \leq 2$

28) $p > 2$

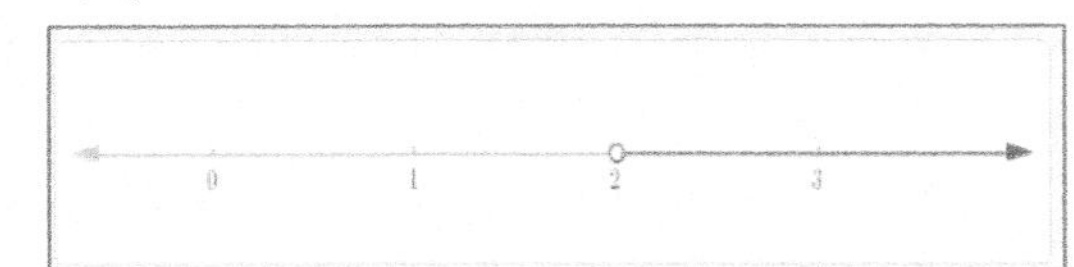

29) $\{(x, y) | y \in R, x \geq \frac{4y+5}{8}\}$

30) $\{(x, y) | y \in R, x \geq \frac{5y+3}{3}\}$

31) $\{(x, y) | y \in R, x \geq \frac{2y-8}{3}\}$

32) $\{(x, y) | y \in R, x \leq \frac{2y+10}{5}\}$

33) $x \geq -\frac{2}{9}$

34) $x < -\frac{4}{11}$

35) $-21 < x < 21$

36) $x > 17$ or $x < -1$

37) $x > 6$ or $x < -18$

38) $-33 < x < 23$

39) $x = -2, y = -4$

40) $x = 3, y = -1$

41) $x = -11, y = -14$

42) $x = -5, y = 3$

43) One solution

44) No solution

45) Infinitely solutions

46) No solution

47) One solution

48) One solution

49) $\begin{cases} y = -3x - 7 \\ y = x + 9 \end{cases}$

50) $\begin{cases} 2x + y = 3 \\ -x + 3y = 4 \end{cases}$

51) $x = 2$

52) 10, 40

53) $C = 15h + 150$

54) 10

55) \$17

56) $\begin{cases} y \le -\frac{1}{2}x + \frac{3}{2} \\ y \ge x \\ y \ge -2 \end{cases}$

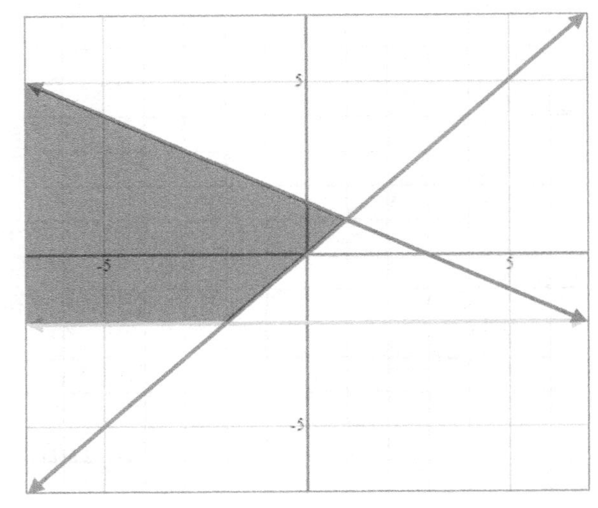

57) $\begin{cases} x < 3 \\ y > -x - 2 \\ y \le x + 1 \end{cases}$

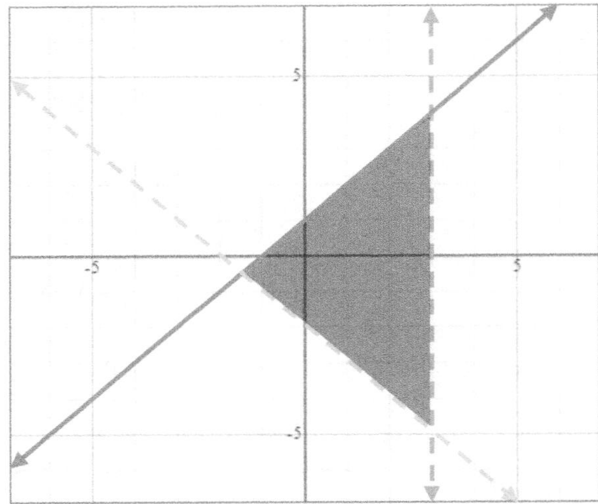

58) $4g + 2 = 38$

59) $3x + y \le 46$

Effortless Math Education

CHAPTER

6 Quadratic

Math topics that you'll learn in this chapter:

- ☑ Solving a Quadratic Equations
- ☑ Graphing Quadratic Functions
- ☑ Solve a Quadratic Equation by Factoring
- ☑ Transformations of Quadratic Functions
- ☑ Quadratic Formula and the Discriminant
- ☑ Characteristics of Quadratic Functions: Equations
- ☑ Characteristics of Quadratic Functions: Graphs
- ☑ Complete a Function Table: Quadratic Functions
- ☑ Domain and Range of Quadratic Functions: Equations
- ☑ Factor Quadratics: Special Cases
- ☑ Factor Quadratics Using Algebra Tiles
- ☑ Write a Quadratic Function from Its Vertex and Another Point

Solving a Quadratic Equations

- Write the equation in the form of: $ax^2 + bx + c = 0$.
- Factorize the quadratic, set each factor equal to zero and solve.
- Use quadratic formula if you can't factorize the quadratic.
- Quadratic formula: $x_{1,2} = \frac{-b \pm \sqrt{b^2 - 4ac}}{2a}$.

Examples:

Find the solutions of each quadratic.

Example 1. $x^2 + 7x + 12 = 0$.

Solution: Factor the quadratic by grouping. We need to find two numbers whose sum is 7 (from $7x$) and whose product is 12. Those numbers are 3 and 4. Then:

$$x^2 + 7x + 12 = 0 \rightarrow x^2 + 3x + 4x + 12 = 0 \rightarrow (x^2 + 3x) + (4x + 12) = 0.$$

Now, find common factors: $(x^2 + 3x) = x(x + 3)$ and $(4x + 12) = 4(x + 3)$. We have two expressions $(x^2 + 3x)$ and $(4x + 12)$ and their common factor is $(x + 3)$. Then: $(x^2 + 3x) + (4x + 12) = 0 \rightarrow x(x + 3) + 4(x + 3) = 0 \rightarrow (x + 3)(x + 4) = 0$.

The product of two expressions is 0. Then:

$$(x + 3) = 0 \rightarrow x = -3 \text{ or } (x + 4) = 0 \rightarrow x = -4.$$

Example 2. $x^2 + 5x + 6 = 0$.

Solution: Use quadratic formula: $x_{1,2} = \frac{-b \pm \sqrt{b^2 - 4ac}}{2a}$, $a = 1$, $b = 5$ and $c = 6$.

Then: $x_{1,2} = \frac{-5 \pm \sqrt{5^2 - 4 \times 1(6)}}{2(1)}$,

$$x_1 = \frac{-5 + \sqrt{5^2 - 4 \times 1(6)}}{2(1)} = -2,\ x_2 = \frac{-5 - \sqrt{5^2 - 4 \times 1(6)}}{2(1)} = -3.$$

Example 3. $x^2 + 6x + 8 = 0$.

Solution: Factor: $x^2 + 6x + 8 = 0 \rightarrow (x + 2)(x + 4) = 0 \rightarrow x = -2$, or $x = -4$.

Graphing Quadratic Functions

- Quadratic functions in vertex form: $y = a(x-h)^2 + k$ where (h, k) is the vertex of the function. The axis of symmetry is $x = h$.
- Quadratic functions in standard form: $y = ax^2 + bx + c$ where $x = -\frac{b}{2a}$ is the value of x in the vertex of the function.
- To graph a quadratic function, first find the vertex, then substitute some values for x and solve for y. (Remember that the graph of a quadratic function is a U −shaped curve and it is called "parabola".)

Example:

Sketch the graph of $y = (x+2)^2 - 3$.

Solution:

Quadratic functions in vertex form:

$$y = a(x-h)^2 + k,$$

and (h, k) is the vertex.

Then, the vertex of $y = (x+2)^2 - 3$ is:

$$(-2, -3).$$

Substitute zero for x and solve for y:

$$y = (0+2)^2 - 3 = 1.$$

The $y-$ intercept is $(0,1)$.

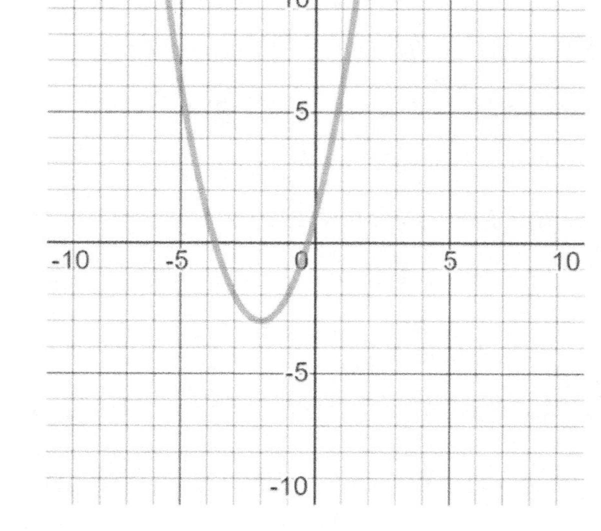

Now, you can simply graph the quadratic function. Notice that quadratic function is a U −shaped curve.

Solve a Quadratic Equation by Factoring

- The general form of a quadratic equation is: $ax^2 + bx + c = 0$.
- When factoring a quadratic equation, we can usually use the multiplication property of zero (MPZ).

  MPZ states that if $p \times q = 0$, then either $p = 0$ or $q = 0$.
- When there are only two terms and they have a common factor, then factoring is relatively simple. This is true of quadratic equations in the form $ax^2 + bx = 0$. (In this equation, the value of c is 0.) The two terms have at least a common factor of x. In this case, first find the greatest common factor (GCF) and factor it out. Then, use the multiplication property of zero (MPZ) to solve the equation.

Examples:

Example 1. Find the solutions of $x^2 - 7x = 0$.

Solution: The greatest common factor of the two terms is x.

Take the common factor out:

$$x(x - 7) = 0.$$

Using MPZ, which states that either $x = 0$ or $x - 7 = 0$.

In the second equation, the value of x equals 7.

$$x - 7 = 0 \rightarrow x = 7.$$

Example 2. Solve $x^2 + 10x - 24 = 0$ by factoring.

Solution: We first factorize the expression:

$$x^2 + 10x - 24 = (x + 12)(x - 2) = 0.$$

Using MPZ, which states that either $(x + 12) = 0$ or $(x - 2) = 0$.

$(x + 12) = 0 \rightarrow x = -12$, $(x - 2) = 0 \rightarrow x = 2$.

bit.ly/3K4phDi
Find more at

Transformations of Quadratic Functions

- In the quadratic equation $y = ax^2$, the graph stretches vertically by the value of unit a. Note that if a is negative, the graph will be inverted.
- In the vertex form, the quadratic function is as follows:

$$f(x) = a(x - h)^2 + k.$$

- In this case, point (h, k) is the vertex of the graph.
- If $k > 0$, the graph moves upwards, and if $k < 0$, the graph moves down.
- If $h > 0$, the graph moves to the right, and if $h < 0$, the graph moves to the left.
- The value of a indicates the elongation of the graph.
  - ❖ If $|a| > 1$, the point corresponding to a certain value x moves farther from the x −axis, so the graph becomes thinner and there will be a vertical elongation.
  - ❖ If $|a| < 1$, the point corresponding to a certain value of x gets closer to the x −axis, so the graph will be wider.

Example:

State the transformations and sketch the graph of the following function.

$y = -3(x + 2)^2 + 4$

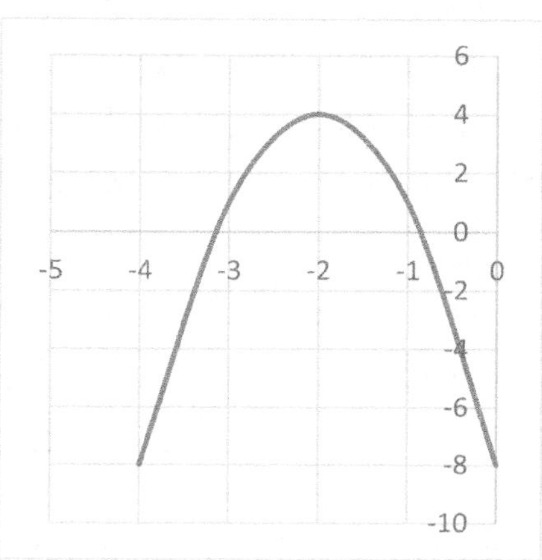

Solution: In this example, since $x - h = x + 2$, then $h = -2$. In this equation, $a = -3$, $h = -2$ and $k = 4$. Since $a < 0$,there is a downward parabola. The vertex is at $(-2,4)$. Find the vertex and some other points to graph the parabola. The negative sign indicates a reflection over the x −axis. The parabola is vertically stretched by a factor of 3. The parabola is shifted 2 units to the left. The parabola is shifted 4 units upward.

$x = 0 \rightarrow y = -3(0 + 2)^2 + 4 = -8,$

$x = -1 \rightarrow y = -3(-1 + 2)^2 + 4 = 1,$

$x = -3 \rightarrow y = -3(-3 + 2)^2 + 4 = 1.$

Find more at bit.ly/3s7MG0o

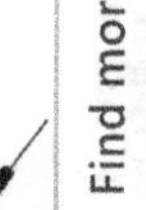

Quadratic Formula and the Discriminant

- A quadratic equation is in the form of $ax^2 + bx + c = 0$. To solve a quadratic equation by the delta (discriminant) method, go through the following steps:

  Step 1: Calculate the delta number as follows: $\Delta = b^2 - 4ac$.

  Step 2: Depending on the value of delta, there are three possible outcomes:

  - If $\Delta < 0$, then the quadratic equation has no real roots.
  - If $\Delta > 0$, then the quadratic equation has two roots, which are obtained by the following formulas:

    $$x_1 = \frac{-b-\sqrt{\Delta}}{2a} \text{ and } x_2 = \frac{-b+\sqrt{\Delta}}{2a}.$$

  - If $\Delta = 0$, then the two roots of the equation are equal and are called double roots:

    $$x_1 = x_2 = \frac{-b}{2a}$$

Examples:

Example 1. Solve equation $5x^2 + 6x + 1 = 0$.

Solution: To solve a quadratic equation, first find the values of a, b, c.
By comparing the mentioned equation with the equation $ax^2 + bx + c = 0$, the values a, b, c are equal to: $a = 5$, $b = 6$, $c = 1$.
Now, calculate delta (Δ). Given the values of a, b, c, the value of Δ is equal to:

$$\Delta = b^2 - 4ac = 6^2 - 4 \times 5 \times 1 = 16.$$

16 is a positive number. Therefore, this equation will have two different solutions:

$$x = \frac{-6 \pm \sqrt{16}}{10} \rightarrow x = \frac{-6 \pm 4}{10} \rightarrow x = -\frac{1}{5}, \text{ or } x = -1.$$

Example 2. Find the solutions of the equation $5x^2 + 2x + 1 = 0$.

Solution: In the equation, the values of a, b, c are: $a = 5$, $b = 2$, $c = 1$.

$$\Delta = b^2 - 4ac = 2^2 - 4 \times 5 \times 1 = -16.$$

The value of delta is negative; Therefore, this equation has no solution in real numbers.

Characteristics of Quadratic Functions: Equations

- Using the equation of a quadratic function, the following can be determined.
  - Direction
  - Vertex
  - Axis of symmetry
  - x −intercept(s)
  - y −intercept
  - range
  - minimum/maximum value

Example:

Specify characteristics of the vertex, direction, and y −intercept for the quadratic function $f(x) = 2x^2 - 3x + 1$.

Solution: In the standard form of a quadratic function $f(x) = ax^2 + bx + c$, the vertex is not immediately obvious. The x −coordinate for the vertex can be obtained by using the formula: $x = -\frac{b}{2a}$,

$$x = -\frac{-3}{2(2)} = \frac{3}{4}$$

Now, substitute it into the equation of function to obtain the y −coordinate:

$$y = f\left(\frac{3}{4}\right) = 2\left(\frac{3}{4}\right)^2 - 3\left(\frac{3}{4}\right) + 1$$

$$= 2\left(\frac{9}{16}\right) - \frac{9}{4} + 1$$

$$= \frac{9}{8} - \frac{9}{4} + 1 = -\frac{1}{8}.$$

So, the vertex of the function $f(x) = 2x^2 - 3x + 1$ is the ordered pair $\left(\frac{3}{4}, -\frac{1}{8}\right)$.

Remember that the sign of the coefficient x^2 indicates the direction of the quadratic equation. Since the coefficient of x^2 is 2, then it is upward.

To find the y −intercept of a function, evaluate the output at $f(0)$.

$$f(0) = 2(0)^2 - 3(0) + 1 = 1.$$

bit.ly/3iRGJU7
Find more at

Characteristics of Quadratic Functions: Graphs

- From the graph of a quadratic function, determine the:
    - vertex
    - axis of symmetry
    - x −intercepts
    - y −intercept
    - domain
    - range
    - minimum/maximum value

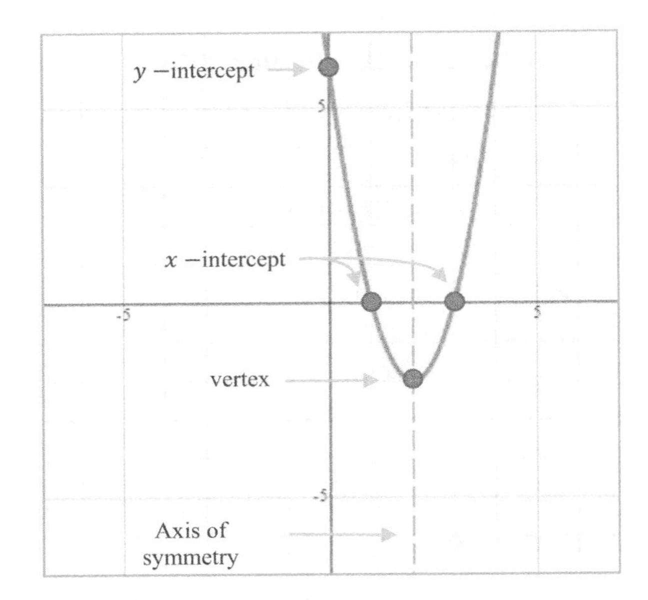

Example:

Considering the following graph, determine the following:

- vertex
- axis of symmetry
- x −intercepts
- y −intercept
- Max/minimum point.

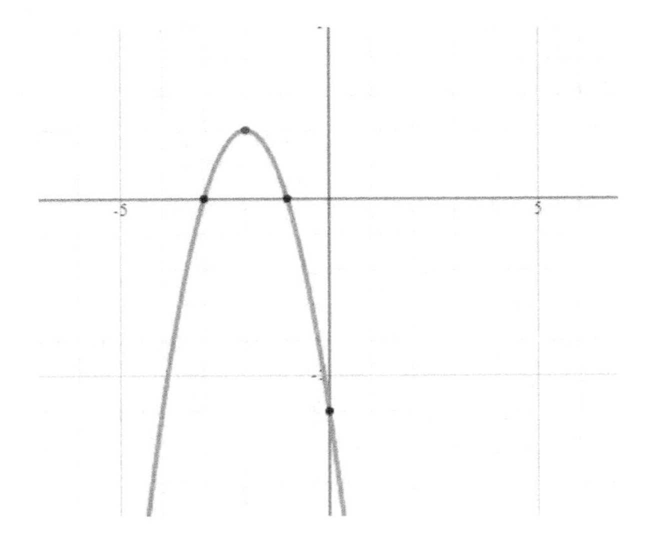

Solution: According to the graph, see that a point at the coordinate $(-2,2)$ is the vertex. So, the line $x = -2$ is the axis of symmetry.

Since the graph intersects the x −axis at points -1 and -3, the mentioned points are x −intercepts. In the same way, the point $(0,-6)$ is the y −intercept.

In addition, this is clear that the vertex is the maximum point.

Complete a Function Table: Quadratic Functions

- To complete a function table of the quadratic function:

  Step 1: Consider the input and output in the table.

  Step 2: Substitute the input into the function.

  Step 3: Evaluate the output.

Example:

Complete the table.

$g(t) = t^2 - 2t + 1$	
t	$g(t)$
-1	
0	
1	

Solution: According to the function table, the first value of the input in the table is -1. Evaluate $g(t) = t^2 - 2t + 1$ for $t = -1$.

$g(t) = t^2 - 2t + 1$	
t	$g(t)$
-1	4
0	
1	

$$g(-1) = (-1)^2 - 2(-1) + 1$$

$$= 1 + 2 + 1 = 4.$$

When $t = -1$, then $g(-1) = 4$. Complete the first row of the table.

In the same way, evaluate $g(t) = t^2 - 2t + 1$ for $t = 0$ and $t = 1$, respectively. So,

$g(t) = t^2 - 2t + 1$	
t	$g(t)$
-1	4
0	1
1	0

$$g(0) = (0)^2 - 2(0) + 1 = 1,$$

$$g(1) = (1)^2 - 2(1) + 1 = 0.$$

Enter the obtained values in the table.

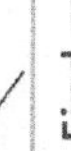

Domain and Range of Quadratic Functions

- The domain of any quadratic function in the form $y = ax^2 + bx + c$ (where a, b, and c are constants) is the set of all real numbers.
- To find the range of a quadratic function, you need to first identify the vertex of the parabola and determine whether it opens upwards or downwards. If the coefficient a is positive ($a > 0$), the parabola opens upwards, and the vertex is the minimum point of the function. Conversely, if a is negative or ($a < 0$) the parabola opens downwards, and the vertex is the maximum point of the function.
- The vertex of a quadratic function can be found using the formula: Vertex $(h, k) = (-\frac{b}{2a}, f(-\frac{b}{2a}))$.

Example:

What is the domain and range of the function to the equation?

$$y = x^2 - 2x + 1.$$

Solution: To find the domain and range of the function $y = x^2 - 2x + 1$, we'll first identify the vertex and determine the direction of the parabola.

The vertex of a quadratic function can be found using the formula:

$$\text{Vertex } (h, k) = (-\frac{b}{2a}, f(-\frac{b}{2a})).$$

In our case, the quadratic function is $y = x^2 - 2x + 1$, where $a = 1$, $b = -2$, and $c = 1$.

$$h = -\frac{(-2)}{2 \times 1} = \frac{2}{2} = 1, k = f(h) = f(1) = (1)^2 - 2(1) + 1 = 1 - 2 + 1 = 0$$

So, the vertex of the parabola is $(1, 0)$.

No, we need to identify the direction of the parabola. Since the coefficient a is positive ($a = 1 > 0$), the parabola opens upwards, and the vertex is the minimum point of the function.

The domain of a quadratic function is all real numbers since it is defined for every x value. Therefore, the domain is: Domain $= \{x \mid x \in R\}$

The range is determined by the direction of the parabola and the vertex. Since the parabola opens upwards, the range will include all values greater than or equal to the y −coordinate (k) of the vertex. In this case, the range will be: Range $= \{y | y \geq 0\}$

Factor Quadratics: Special Cases

- Special cases square:
  - Factor an equation of the form: $a^2 \pm 2ab + b^2 = (a \pm b)^2$
  - Factor an equation of the form: $a^2 - b^2 = (a + b)(a - b)$

Examples:

Example 1. Factor $4x^2 - 36$.

Solution: Notice that $4x^2$ and 36 are perfect squares, because $4x^2 = (2x)^2$, and $36 = 6^2$. By using the formula $a^2 - b^2 = (a + b)(a - b)$. Let $a = 2x$, and $b = 6$. The represented equation can be rewritten as follows:

$$4x^2 - 36 = (2x + 6)(2x - 6).$$

Example 2. Factor $4k^2 - 4k + 1$.

Solution: First, consider that this formula $a^2 - 2ab + b^2 = (a - b)^2$. Since $4k^2 = (2k)^2$, then $4k^2$ and 1 are perfect squares. This means that $a = 2k$, and $b = 1$. Next, check to see if the middle term is equal to $2ab$, which it is: $2ab = 2(2k)1 = 4k$. Therefore, the square equation can be rewritten as,

$$4k^2 - 4k + 1 = (2k - 1)^2.$$

Example 3. Factor $36x^2 + 36x + 9$.

Solution: First, notice that $36x^2$ and 9 are perfect squares because $36x^2 = (6x)^2$ and $9 = 3^2$. Let $a = 6x$, and $b = 3$. Now, evaluate the middle term $36x$. Then, you can write as $36x = 2(6x)3 = 2ab$. Using this formula: $a^2 + 2ab + b^2 = (a + b)^2$. Therefore, we have:

$$36x^2 + 36x + 9 = (6x + 3)^2.$$

Example 4. Factor $2x^2 - 16x + 32$.

Solution: First, rewrite the equation as $2x^2 - 16x + 32 = 2(x^2 - 8x + 16)$. According to the x^2 and 16 are perfect squares because $x^2 = (x)^2$ and $16 = 4^2$. That is, $a = x$, and $b = 4$. Now, check to see if the middle term is equal to $2ab$: $2ab = 2(x)4 = 8x$. Using this formula: $a^2 + 2ab + b^2 = (a + b)^2$.
Therefore, we have: $2x^2 - 16x + 32 = 2(x^2 - 8x + 16) = 2(x - 4)^2$.

Factor Quadratics Using Algebra Tiles

- To factor quadratic expressions like $ax^2 + bx + c$ using algebraic tiles, follow the steps below:
  - Model the polynomials with tiles.
  - Arrange the tiles into a rectangle grid. Start with the x^2 tiles from the upper left corner so that the number of horizontal and vertical divisions is equal to the multiples of a. Add the integer tiles in the lower right corner. Here, the number of horizontal and vertical divisions should be equal to integer multiples. Make sure to choose horizontal and vertical divisions from possible multiples of a and c that will fill the remaining empty grid tiles.
  - The product of expressions related to horizontal and vertical divisions is equal to the answer.
  - Check the answer.

Example:

Use algebra tiles to factor: $3x^2 - 7x + 2$.

Solution: Model the polynomials with tiles:
In this case, arrange the tiles into a rectangle grid.

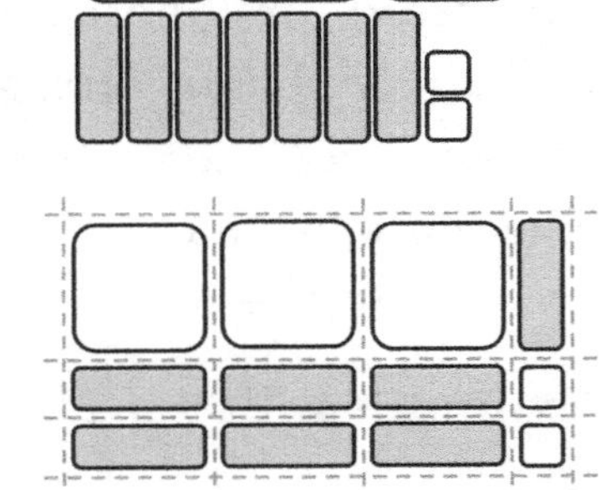

Determine both binomials relate to the divisions, such that $3x - 1$ for the horizontal division and $x - 2$ for the vertical. As follow:

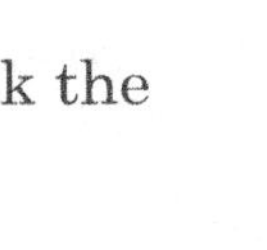

In the end, multiply two expressions and check the answer,

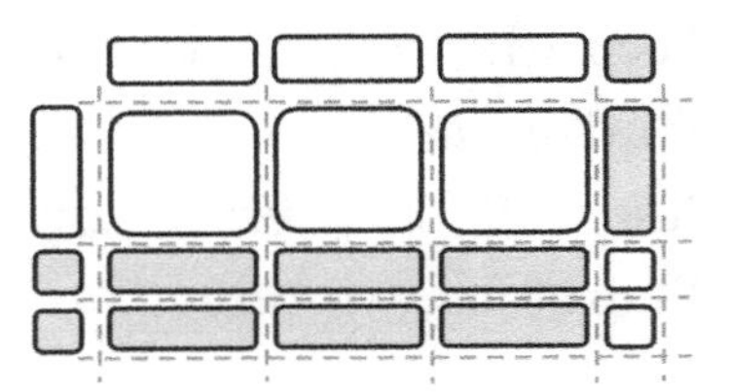

$$(3x - 1)(x - 2) = 3x^2 - 7x + 2.$$

Write a Quadratic Function from Its Vertex and Another Point

- A quadratic function in the standard form of $y = ax^2 + bx + c$ can be represented in the vertex form as follows:

  $$y = a(x-h)^2 + k,$$

  where (h, k) is the coordinate of the vertex.

- So that if $a > 0$, the function opens up and the coordinate vertex is the minimum value of the function. If $a < 0$, then the function is downward, and the coordinate vertex is the maximum value.

Examples:

Example 1. A quadratic function opening up or down has vertex (0,1) and passes through (2,0). Write its equation in vertex form.

Solution: Use the vertex form of the quadratic function as $y = a(x-h)^2 + k$. Put the coordinate of the vertex (0,1) in the vertex form:

$$(0,1) \rightarrow 1 = a(x-0)^2 + 1 \rightarrow y = ax^2 + 1.$$

To find a, substitute (2,0) in this equation and calculate. Then,

$$(2,0) \rightarrow 0 = a(2)^2 + 1 \rightarrow a = -\frac{1}{4}.$$

Therefore, the equation of the quadratic function in the vertex form is as follows:

$$y = -\frac{1}{4}x^2 + 1.$$

Example 2. A quadratic function has vertex (0,0) and passes through $(-12, -18)$. Write its equation in vertex form.

Solution: By using the vertex form formula: $y = a(x-h)^2 + k$. So, we have:

$$(0,0) \rightarrow y = a(x-0)^2 + 0 \rightarrow y = ax^2.$$

Substitute $(-12, -18)$ in the obtained equation, then:

$$-18 = a(-12)^2 \rightarrow -18 = 144a \rightarrow a = -\frac{18}{144} = -\frac{1}{8}.$$

Therefore, $y = -\frac{1}{8}x^2$.

Chapter 6: Practices

 **Solve each equation by factoring or using the quadratic formula.**

1) $x^2 - x - 2 = 0$
2) $x^2 - 6x + 8 = 0$
3) $x^2 - 4x + 3 = 0$
4) $x^2 + x - 12 = 0$
5) $x^2 + 7x - 18 = 0$
6) $x^2 - 2x - 15 = 0$
7) $x^2 + 6x - 40 = 0$
8) $x^2 - 9x - 36 = 0$

 **Sketch the graph of each function.**

9) $y = (x - 4)^2 - 2$

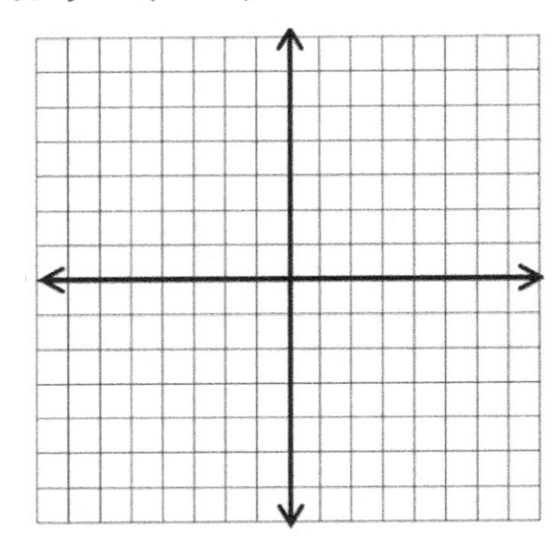

10) $y = 2(x + 2)^2 - 3$

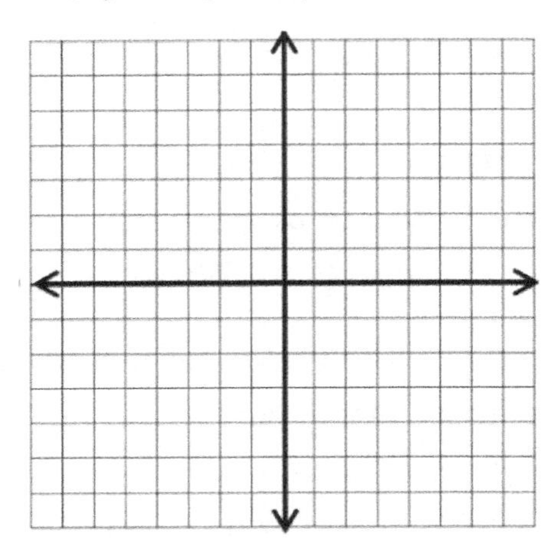

 **Solve each equation by factoring or using the quadratic formula.**

11) $x^2 - 2x - 3 = 0$
12) $x^2 + 9x + 20 = 0$

State the transformations and sketch the graph of the following function.

13) $y = 2(x - 3)^2 + 1$

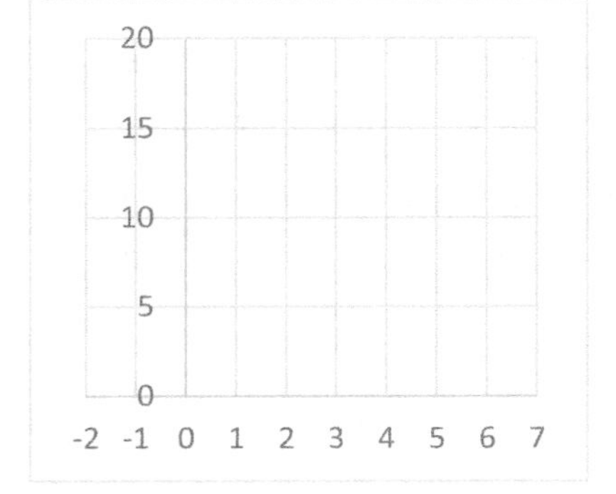

**Find the answer to the equation.**

14) $2x^2 - 7x + 3 = 0$

15) $x^2 + 8x - 9 = 0$

16) $2x^2 + 5x - 3 = 0$

17) $x^2 + 6x + 9 = 0$

 **Solve.**

18) Find the equation of the axis of symmetry for the parabola $y = x^2 + 7x + 3$.

19) Find the y-intercept of the parabola $x^2 + 25x + 7$.

20) Find the vertex of the parabola $y = x^2 - 4x + 3$.

 **Considering the following graph, determine the following:**

21) vertex

22) axis of symmetry

23) y –intercepts

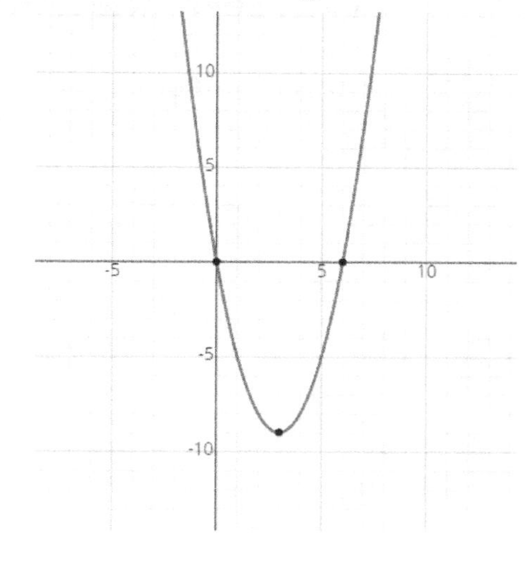

 **Complete the table.**

24)

$g(t) = t^2 + 7$	
t	$g(t)$
-1	
0	
1	

25)

$f(p) = 4p^2$	
p	$f(p)$
-2	
0	
2	

Determine the domain and range of each function.

26) $y = x^2 + 5x + 6$

27) $y = x^2 + 3$

28) $y = -x^2 + 4$

Factor.

29) $25x^2 + 20x + 4$

30) $9x^2 - 1$

31) $3 + 6x + 3x^2$

32) $b^4 - 36$

Use algebra tiles to factor.

33) $x^2 - 3x + 2$

34) $x^2 + 5x + 6$

Write each quadratic function as a vertex form.

35) A parabola opening or down has vertex $(0,0)$ and passes through $(8, -16)$.

36) A parabola opening up or down has vertex $(0, 2)$ and passes through $(-2,5)$.

Chapter 6: Answers

1) $x = 2, x = -1$
2) $x = 2, x = 4$
3) $x = 3, x = 1$
4) $x = 3, x = -4$
5) $x = 2, x = -9$
6) $x = 5, x = -3$
7) $x = 4, x = -10$
8) $x = 12, x = -3$
9) $y = (x - 4)^2 - 2$

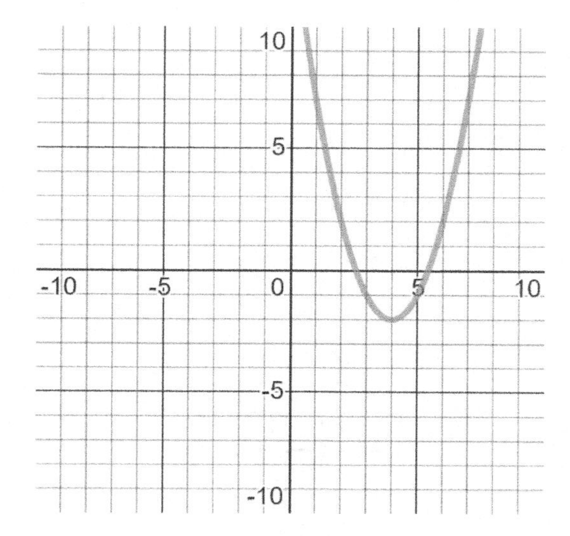

10) $y = 2(x + 2)^2 - 3$

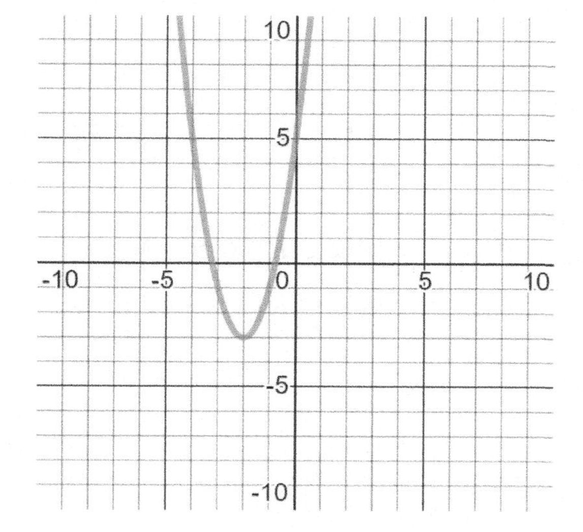

11) $\{3, -1\}$
12) $\{-4, -5\}$
13) The graph stretches vertically by a factor of 2.
Move 3 units to the right and 1 unit up.

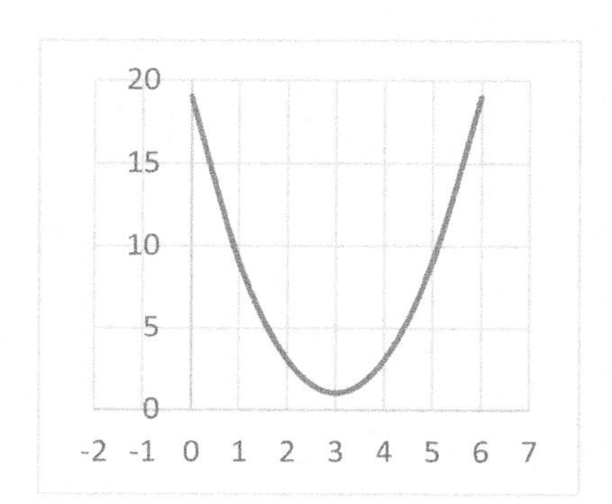

14) $x_1 = 3, x_2 = \frac{1}{2}$
15) $x_1 = -9, x_2 = 1$
16) $x_1 = -3, x_2 = \frac{1}{2}$
17) $x_1 = x_2 = -3$
18) $x = -\frac{7}{2}$
19) 7

20) $(2, -1)$

21) $(3, -9)$

22) 3

23) 0

24)

$g(t) = t^2 + 7$	
t	$g(t)$
-1	8
0	7
1	8

25)

$f(p) = 4p^2$	
p	$f(p)$
-2	16
0	0
2	16

26) $D = \{x|x \in R\}, R = \{y \in R|y \geq -0.25\}$

27) $D = \{x|x \in R\}, R = \{R|y \geq 3\}$

28) $D = \{x|x \in R\}, R = \{y \in R|y \leq 4\}$

29) $(5x + 2)^2$

30) $(3x - 1)(3x + 1)$

31) $3(1 + x)^2$

32) $(b^2 + 6)(b^2 - 6)$

33) $(x - 1)(x - 2)$

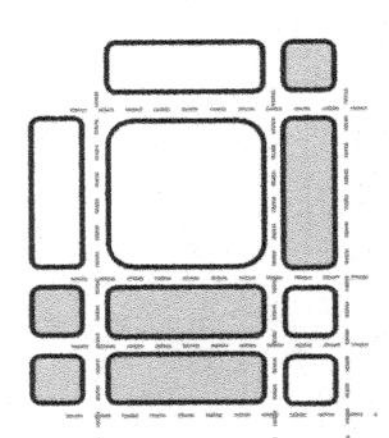

34) $(x + 2)(x + 3)$

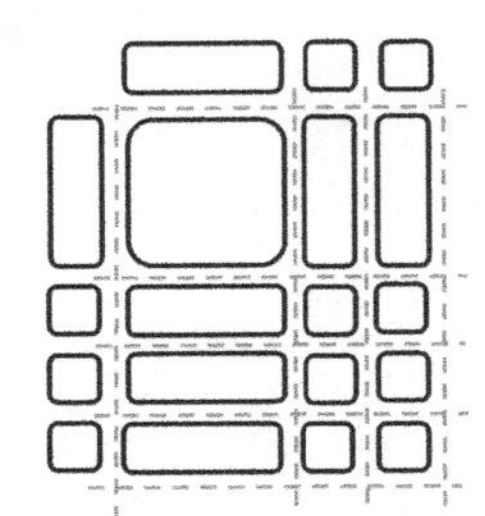

35) $y = -\frac{1}{4}x^2$

36) $y = \frac{3}{4}x^2 + 2$

CHAPTER

7 Polynomials

Math topics that you'll learn in this chapter:

- ☑ Simplifying Polynomials
- ☑ Adding and Subtracting Polynomials
- ☑ Add and Subtract Polynomials Using Algebra Tiles
- ☑ Multiplying Monomials
- ☑ Multiplying and Dividing Monomials
- ☑ Multiplying a Polynomial and a Monomial
- ☑ Multiply Polynomials Using Area Models
- ☑ Multiplying Binomials
- ☑ Multiply two Binomials Using Algebra Tiles
- ☑ Factoring Trinomials
- ☑ Factoring Polynomials
- ☑ Use a Graph to Factor Polynomials
- ☑ Factoring Special Case Polynomials
- ☑ Add Polynomials to Find Perimeter

Simplifying Polynomials

- To simplify polynomials, find "like" terms. (They have the same variables with the same power).
- Use "FOIL". (First–Out–In–Last) for binomials:

$$(x + a)(x + b) = x^2 + (b + a)x + ab$$

- Add or subtract "like" terms using order of operation.

Examples:

Example 1. Simplify this expression. $x(4x + 7) - 2x =$

Solution: Use the Distributive Property:

$$x(4x + 7) = 4x^2 + 7x.$$

Now, combine like terms:

$$x(4x + 7) - 2x = 4x^2 + 7x - 2x = 4x^2 + 5x.$$

Example 2. Simplify this expression. $(x + 3)(x + 5) =$

Solution: First, apply the FOIL method:

$$(a + b)(c + d) = ac + ad + bc + bd.$$

Therefore:

$$(x + 3)(x + 5) = x^2 + 5x + 3x + 15.$$

Now combine like terms:

$$x^2 + 5x + 3x + 15 = x^2 + 8x + 15.$$

Example 3. Simplify this expression. $2x(x - 5) - 3x^2 + 6x =$.

Solution: Use the Distributive Property:

$$2x(x - 5) = 2x^2 - 10x.$$

Then:

$$2x(x - 5) - 3x^2 + 6x = 2x^2 - 10x - 3x^2 + 6x.$$

Now combine like terms:

$$2x^2 - 3x^2 = -x^2, \text{ and } -10x + 6x = -4x.$$

The simplified form of the expression:

$$2x^2 - 10x - 3x^2 + 6x = -x^2 - 4x.$$

Adding and Subtracting Polynomials

- Adding polynomials is just a matter of combining like terms, with some order of operations considerations thrown in.
- Be careful with the minus signs, and don't confuse addition and multiplication!
- For subtracting polynomials, sometimes you need to use the Distributive Property: $a(b + c) = ab + ac$, $a(b - c) = ab - ac$.

Examples:

Example 1. Simplify the expressions. $(x^2 - 2x^3) - (x^3 - 3x^2) =$

Solution: First, use the Distributive Property: $-(x^3 - 3x^2) = -x^3 + 3x^2$.
$\rightarrow (x^2 - 2x^3) - (x^3 - 3x^2) = x^2 - 2x^3 - x^3 + 3x^2$.
Now combine like terms: $-2x^3 - x^3 = -3x^3$ and $x^2 + 3x^2 = 4x^2$.
Then: $(x^2 - 2x^3) - (x^3 - 3x^2) = x^2 - 2x^3 - x^3 + 3x^2 = -3x^3 + 4x^2$

Example 2. Add expressions. $(3x^3 - 5) + (4x^3 - 2x^2) =$

Solution: Remove parentheses:
$(3x^3 - 5) + (4x^3 - 2x^2) = 3x^3 - 5 + 4x^3 - 2x^2$.
Now combine like terms: $3x^3 - 5 + 4x^3 - 2x^2 = 7x^3 - 2x^2 - 5$.

Example 3. Simplify the expressions. $(-4x^2 - 2x^3) - (5x^2 + 2x^3) =$

Solution: First, use the Distributive Property: $-(5x^2 + 2x^3) = -5x^2 - 2x^3 \rightarrow$
$(-4x^2 - 2x^3) - (5x^2 + 2x^3) = -4x^2 - 2x^3 - 5x^2 - 2x^3$.
Now combine like terms and write in standard form:
$-4x^2 - 2x^3 - 5x^2 - 2x^3 = -4x^3 - 9x^2$.

Example 4. Simplify the expressions. $(8x^2 - 3x^3) + (2x^2 + 5x^3) =$

Solution: Remove parentheses:

$(8x^2 - 3x^3) + (2x^2 + 5x^3) = 8x^2 - 3x^3 + 2x^2 + 5x^3$.

Now combine like terms and write in standard form:

$8x^2 - 3x^3 + 2x^2 + 5x^3 = 10x^2 + 2x^3 = 2x^3 + 10x^2$.

Add and Subtract Polynomials Using Algebra Tiles

- To better understand and visualize the addition and subtraction of algebraic expressions, you can use Algebra tiles as follow:
  - Model the polynomials using tiles.
  - For algebraic subtraction, change the color of the tiles on the second side.
  - Cross out the same number of negative or positive tiles on both sides of the equation.
  - Write the answer by determining the number of remaining tiles.

Examples

Example1. Use algebra tiles to simplify: $(x^2 - x + 3) + (2x^2 + 3x - 2)$.

Solution: Model the given polynomials using algebra tiles.

Here, cross out one x tile on the left side and do the same on the other side. In the same way, cancel two 1 tiles on the left side and do the same on the right side. That is,

Count the number of remaining tiles. So, $3x^2 + 2x + 1$.

Example2. Simplify the polynomial $(2x^2 + 3x - 1) - (x^2 - 2x - 2)$ using algebra tiles.

Solution: Model the polynomials with tiles.

Change the color of the tiles on the second side, then add them to the first side.

Now, simplify the obtained algebraic tiles by canceling negative and positive tiles of the same size. As follow:

Finally, by counting the remaining tiles, the following expression obtains: $x^2 + 5x + 1$.

Multiplying Monomials

- A monomial is a polynomial with just one term: Examples: $2x$ or $7y^2$.
- When you multiply monomials, first multiply the coefficients (a number placed before) and then multiply the variables using multiplication property of exponents.

$$x^a \times x^b = x^{a+b}$$

Examples:

Example 1. Multiply expressions. $2xy^3 \times 6x^4y^2$.

Solution: Find the same variables and use the multiplication property of exponents: $x^a \times x^b = x^{a+b}$.
$x \times x^4 = x^{1+4} = x^5$ and $y^3 \times y^2 = y^{3+2} = y^5$.
Then, multiply coefficients and variables: $2xy^3 \times 6x^4y^2 = 12x^5y^5$.

Example 2. Multiply expressions. $7a^3b^8 \times 3a^6b^4 =$

Solution: Use the multiplication property of exponents: $x^a \times x^b = x^{a+b}$.
$a^3 \times a^6 = a^{3+6} = a^9$ and $b^8 \times b^4 = b^{8+4} = b^{12}$.
Then: $7a^3b^8 \times 3a^6b^4 = 21a^9b^{12}$.

Example 3. Multiply. $5x^2y^4z^3 \times 4x^4y^7z^5$

Solution: Use the multiplication property of exponents: $x^a \times x^b = x^{a+b}$.
$x^2 \times x^4 = x^{2+4} = x^6$, $y^4 \times y^7 = y^{4+7} = y^{11}$ and $z^3 \times z^5 = z^{3+5} = z^8$.
Then: $5x^2y^4z^3 \times 4x^4y^7z^5 = 20x^6y^{11}z^8$.

Example 4. Simplify. $(-6a^7b^4)(4a^8b^5) =$

Solution: Use the multiplication property of exponents: $x^a \times x^b = x^{a+b}$.
$a^7 \times a^8 = a^{7+8} = a^{15}$ and $b^4 \times b^5 = b^{4+5} = b^9$.
Then: $(-6a^7b^4)(4a^8b^5) = -24a^{15}b^9$.

Multiplying and Dividing Monomials

- When you divide or multiply two monomials, you need to divide or multiply their coefficients and then divide or multiply their variables.
- In case of exponents with the same base, for Division, subtract their powers, for Multiplication, add their powers.
- Exponent's Multiplication and Division rules:

$$x^a \times x^b = x^{a+b}, \frac{x^a}{x^b} = x^{a-b}$$

Examples:

Example 1. Multiply expressions. $(3x^5)(9x^4) =$

Solution: Use the multiplication property of exponents:
$x^a \times x^b = x^{a+b} \rightarrow x^5 \times x^4 = x^9$, then: $(3x^5)(9x^4) = 27x^9$.

Example 2. Multiply expressions. $(-5x^8)(4x^6) =$

Solution: Use the multiplication property of exponents:

$x^a \times x^b = x^{a+b} \rightarrow x^8 \times x^6 = x^{14}$.

Then: $(-5x^8)(4x^6) = -20x^{14}$.

Example 3. Divide expressions. $\frac{12x^4y^6}{6xy^2} =$

Solution: Use the division property of exponents:
$\frac{x^a}{x^b} = x^{a-b} \rightarrow \frac{x^4}{x} = x^{4-1} = x^3$ and $\frac{y^6}{y^2} = y^{6-2} = y^4$.
Then: $\frac{12x^4y^6}{6xy^2} = 2x^3y^4$.

Example 4. Divide expressions. $\frac{49a^6b^9}{7a^3b^4}$

Solution: Use the division property of exponents:
$\frac{x^a}{x^b} = x^{a-b} \rightarrow \frac{a^6}{a^3} = a^{6-3} = a^3$ and $\frac{b^9}{b^4} = b^{9-4} = b^5$.
Then: $\frac{49a^6b^9}{7a^3b^4} = 7a^3b^5$.

Multiplying a Polynomial and a Monomial

- When multiplying monomials, use the product rule for exponents.

$$x^a \times x^b = x^{a+b}$$

- When multiplying a monomial by a polynomial, use the distributive property.

$$a \times (b + c) = a \times b + a \times c = ab + ac$$

$$a \times (b - c) = a \times b - a \times c = ab - ac$$

Examples:

Example 1. Multiply expressions. $6x(2x + 5)$

Solution: Use the Distributive Property:

$6x(2x + 5) = 6x \times 2x + 6x \times 5 = 12x^2 + 30x$.

Example 2. Multiply expressions. $x(3x^2 + 4y^2)$

Solution: Use the Distributive Property:

$x(3x^2 + 4y^2) = x \times 3x^2 + x \times 4y^2 = 3x^3 + 4xy^2$.

Example 3. Multiply. $-x(-2x^2 + 4x + 5)$

Solution: Use the Distributive Property:

$-x(-2x^2 + 4x + 5) = (-x) \times (-2x^2) + (-x) \times (4x) + (-x) \times (5) = 2x^3 - 4x^2 - 5x$.

Example 4. Multiply. $-4x(-5x^2 + 3x - 6)$

Solution: Use the Distributive Property:

$-4x(-5x^2 + 3x - 6) = (-4x) \times (-5x^2) + (-4x) \times (3x) +$

$(-4x) \times (-6) = 20x^3 - 12x^2 + 24x$.

bit.ly/3aBYdx2
Find more at

Multiply Polynomials Using Area Models

- To multiply polynomials using area models, follow the steps below:
  - Model a rectangular area such that each side corresponds to the polynomials.
  - Divide the sides associated with each of the polynomials. Separate each polynomial side into the monomial factors.
  - Complete the areas by multiplying the monomials.
  - To find the product of polynomials, add the resulting expressions.

Examples:

Example 1. Use the area model to find the product $2x(x+1)$.

Solution: Model a rectangular area,

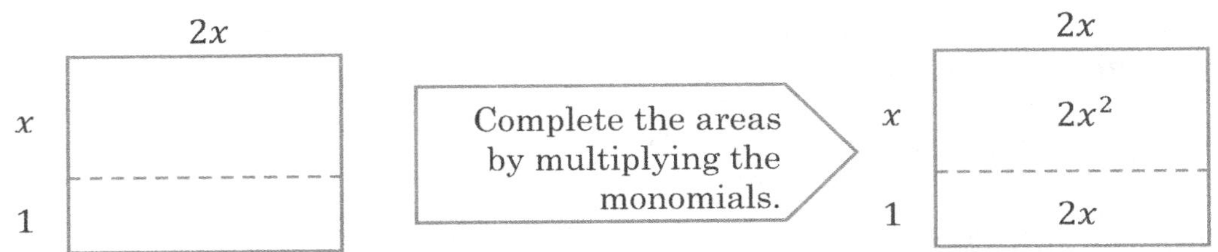

Last, combine terms to find the polynomial product.

$2x(x+1) = 2x^2 + 2x$.

Example 2. Use an area model to multiply these binomials.

$$(a-2)(3a+1)$$

Solution: Draw an area model representing the product $(a-2)(3a+1)$.

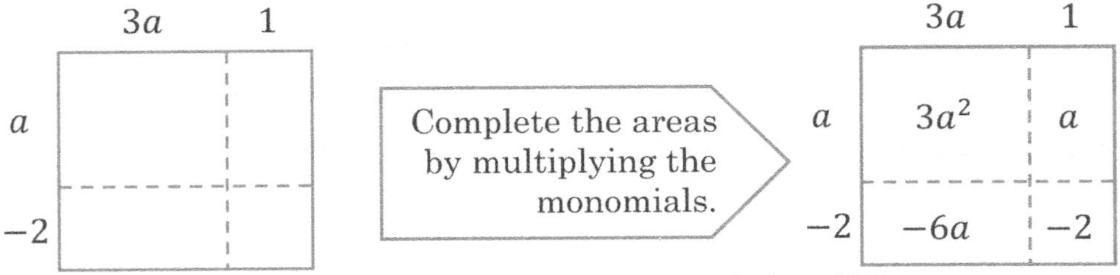

Now, add the partial products to find the product and simplify,

$3a^2 + a - 6a - 2 = 3a^2 - 5a - 2$.

Therefore, $(a-2)(3a+1) = 3a^2 - 5a - 2$.

Multiplying Binomials

- A binomial is a polynomial that is the sum or the difference of two terms, each of which is a monomial.
- To multiply two binomials, use the "FOIL" method. (First–Out–In–Last)

$$(x + a)(x + b) = x \times x + x \times b + a \times x + a \times b = x^2 + bx + ax + ab$$

Examples:

Example 1. Multiply Binomials. $(x + 3)(x - 2) =$

Solution: Use the "FOIL". (First–Out–In–Last):

$(x + 3)(x - 2) = x^2 - 2x + 3x - 6.$

Then combine like terms: $x^2 - 2x + 3x - 6 = x^2 + x - 6.$

Example 2. Multiply. $(x + 6)(x + 4) =$

Solution: Use the "FOIL". (First–Out–In–Last):

$(x + 6)(x + 4) = x^2 + 4x + 6x + 24.$

Then simplify: $x^2 + 4x + 6x + 24 = x^2 + 10x + 24.$

Example 3. Multiply. $(x + 5)(x - 7) =$

Solution: Use the "FOIL". (First–Out–In–Last):

$(x + 5)(x - 7) = x^2 - 7x + 5x - 35.$

Then simplify: $x^2 - 7x + 5x - 35 = x^2 - 2x - 35.$

Example 4. Multiply Binomials. $(x - 9)(x - 5) =$

Solution: Use the "FOIL". (First–Out–In–Last):

$(x - 9)(x - 5) = x^2 - 5x - 9x + 45.$

Then combine like terms: $x^2 - 5x - 9x + 45 = x^2 - 14x + 45.$

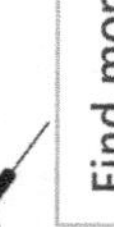

Multiply two Binomials Using Algebra Tiles

- To find the product of two binomials using algebraic tiles, do these in the following way:
  - Set up the grid so that the horizontal divisions correspond to one of the binomials and the vertical divisions to the other side.
  - Match the appropriate tiles in this grid.
  - To specify the answer, sum the like terms inside the grid.
- Remember these rules for performing the multiplication of binomials containing a negative term:
  - Two positives or two negatives equal a positive value.
  - A positive and a negative equal a negative value.

Example:

Use algebra tiles to simplify: $(x - 2)(2x + 1)$.

Solution: Set up the grid as follow:

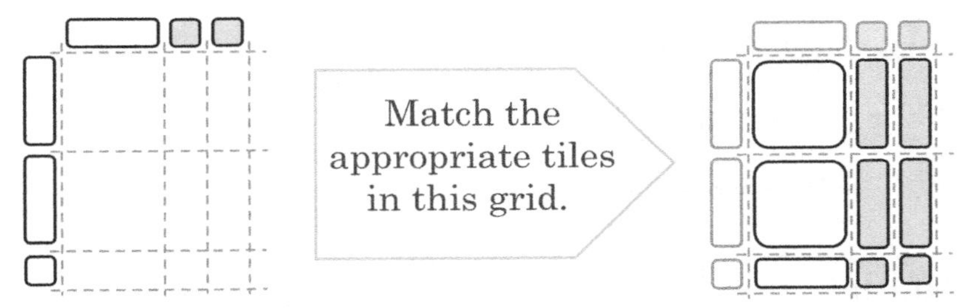

Here, cross out one x tile on the first column and do the same on the second column.

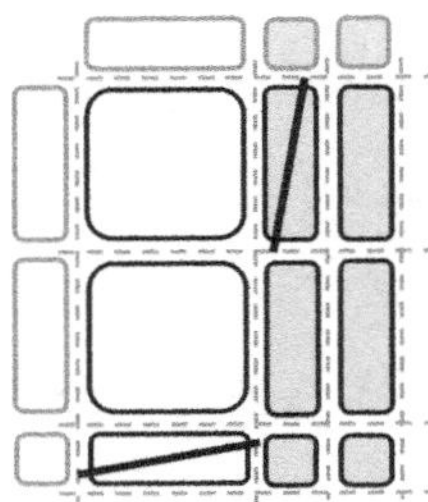

Count the like terms inside the grid. Since the number of x^2 tiles $= 2$, the number of $-x$ tiles $= 3$, and the number of -1 tiles $= 2$, then, sum the like terms inside the grid. So, $2x^2 - 3x - 2$.

Factoring Trinomials

- To factor trinomials, you can use following methods:
  - "FOIL": $(x + a)(x + b) = x^2 + (b + a)x + ab$.
  - "Difference of Squares":

$$a^2 - b^2 = (a + b)(a - b)$$
$$a^2 + 2ab + b^2 = (a + b)(a + b)$$
$$a^2 - 2ab + b^2 = (a - b)(a - b)$$

  - "Reverse FOIL": $x^2 + (b + a)x + ab = (x + a)(x + b)$.

Examples:

Example 1. Factor this trinomial. $x^2 - 2x - 8$

Solution: Break the expression into groups. You need to find two numbers that their product is -8 and their sum is -2. (remember "Reverse FOIL": $x^2 + (b + a)x + ab = (x + a)(x + b)$). Those two numbers are 2 and -4. Then: $x^2 - 2x - 8 = (x^2 + 2x) + (-4x - 8)$.

Now factor out x from $x^2 + 2x$: $x(x + 2)$, and factor out -4 from $-4x - 8$: $-4(x + 2)$; then: $(x^2 + 2x) + (-4x - 8) = x(x + 2) - 4(x + 2)$

Now factor out like term: $(x + 2)$. Then: $(x + 2)(x - 4)$.

Example 2. Factor this trinomial. $x^2 - 2x - 24$

Solution: Break the expression into groups: $(x^2 + 4x) + (-6x - 24)$.

Now factor out x from $x^2 + 4x$: $x(x + 4)$, and factor out -6 from $-6x - 24$: $-6(x + 4)$; then: $x(x + 4) - 6(x + 4)$, now factor out like term: $(x + 4) \rightarrow x(x + 4) - 6(x + 4) = (x + 4)(x - 6)$.

Factoring Polynomials

- To factor a polynomial:
  - Step 1: Break down each term into its prime factors.
  - Step 2: Find GCF (greatest common factor).
  - Step 3: Factor the GCF out from each term.
  - Step 4: Simplify as needed.
  - To factor a polynomial, you can also use these formulas:

$$(x+a)(x+b) = x^2 + (b+a)x + ab$$

$$a^2 - b^2 = (a+b)(a-b)$$

$$a^2 + 2ab + b^2 = (a+b)(a+b)$$

$$a^2 - 2ab + b^2 = (a-b)(a-b)$$

Examples:

Factor each polynomial.

Example 1. $6x^3 - 6x$

Solution: To factorize the expression $6x^3 - 6x$, first factor out the largest common factor, $6x$, and then you will see that you have the pattern of the difference between two complete squares: $(a-b)(a+b) = a^2 - b^2$, then: $6x^3 - 6x = 6x(x^2 - 1) = 6x(x-1)(x+1)$.

Example 2. $x^2 + 9x + 20$

Solution: To factorize the expression $x^2 + 9x + 20$, you need to find two numbers whose product is 20 and sum is 9. You can get the number 20 by multiplying 1×20, 2×10, 4×5.The last pair will be your choice, because $4 + 5 = 9$.

Then: $x^2 + 9x + 20 = (x+4)(x+5)$.

Use a Graph to Factor Polynomials

- If a polynomial includes the factor of the form $(x-h)^p$, you can determine the behavior near the x –intercept by the power p. It can be said that $x = h$ is a zero of multiplicity p:
  - If a polynomial function graph touches the x –axis, it's zero with even multiplicity.
  - If a polynomial function' graph crosses the x –axis, it's a zero with odd multiplicity.
  - A polynomial function' graph gets flattered at zero if the multiplicity of the zero is higher.
  - The sum of the multiplicities of the zero is the polynomial function's degree.
- To check factorization using a graphing calculator, follow these steps:
  - 1st step: Press the $Y =$ button and enter the given equation for $Y1$.
  - 2nd step: Press the GRAPH button to see the equation's graph.
  - 3rd step: Press the TRACE button and by the left and right buttons move the cursor along the graph. You can see at which points the graph crosses the x –axis
  - 4th step: To find the y –values at the x –values when the graph crosses the x –axis, enter the value of x at this point and press ENTER button while in Trace mode. The calculator finds the y – value for you. The calculator tells you that in these values of x the y –values are equal to zero.
  - 5th step: Remember that for functions with binomial factors of the form $(x-a)$, a is an x –intercept.

Example:

Use a graph to factor following polynomial.

$$x^2 - x - 2$$

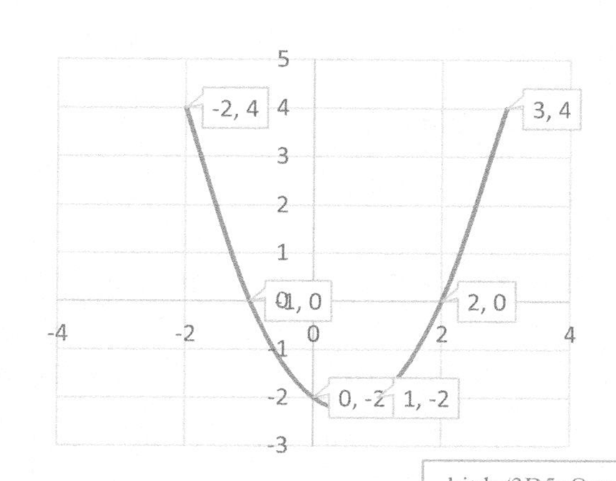

Solution: First, graph the polynomial. Then find the points where the polynomial function graph crosses the x –axis. These points are the zeros of the polynomial function. For $x^2 - x - 2, x = -1$ and $x = 2$ are the zeros of the polynomial function.

bit.ly/3D5vQom
Find more at

Factoring Special Case Polynomials

- There are different methods to factor a special polynomial. Here are some common polynomial cases to factor:
- To solve the difference between two complete squares:

$$x^2 - a^2 = (x - a)(x + a)$$

- To solve a perfect square trinomial:

$$a^2 + 2ab + b^2 = (a + b)^2$$

$$a^2 - 2ab + b^2 = (a - b)^2$$

- FOIL:

$$(x + a)(x + b) = x^2 + (b + a)x + ab$$

- Reverse FOIL:

$$x^2 + (b + a)x + ab = (x + a)(x + b)$$

Examples:

Example 1. Factor completely. $25y^2 + 30y + 9$

Solution: You may notice that two terms $25y^2$ and 9 are perfect squares. The root $25y^2$ is equal to $5y$, and the root 9 is equal to 3. The middle expression, $30y$, is equal to twice the product of $5y$ and 3. So, you have a perfect square trinomial whose factoring result is $(5y + 3)^2$.

$25y^2 + 30y + 9 = (5y + 3)^2$.

Example 2. Factor completely. $36y^2 - 25x^4$

Solution: Phrase $36y^2$ can be written in form $(6y)^2$ and phrase $25x^4$ in form $(5x^2)^2$. Therefore, the relation of the question form is as follows:

$(6y)^2 - (5x^2)^2 = (6y - 5x^2)(6y + 5x^2)$.

Add Polynomials to Find Perimeter

- To find the perimeter of a two-dimensional shape whose sides are given as polynomials, the sum of polynomials is calculated.

Examples

Example 1. Find the perimeter. Simplify your answer.

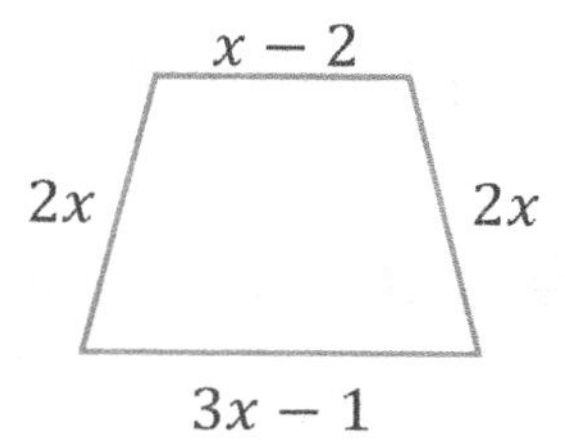

Solution: The perimeter of the shape is the sum of the sides. So,

$$\text{Perimeter} = (x - 2) + (2x) + (3x - 1) + (2x)$$
$$= x - 2 + 2x + 3x - 1 + 2x.$$

Group and add like terms,

$$\text{Perimeter} = (x + 2x + 3x + 2x) + (-2 - 1)$$
$$= 8x - 3.$$

Example 2. What is the perimeter of the rectangle if the length is $2x^2 - 1$ and the width is $4x + 2$?

$2x^2 - 1$

$4(x + 1)$

Solution: The perimeter of the rectangle is,

$$\text{Perimeter} = 2\big((2x^2 - 1) + 4(x + 1)\big)$$

Expand the expression and simplify,

$$\text{Perimeter} = 2\big((2x^2 - 1) + (4x + 4)\big)$$
$$= 2(2x^2 - 1 + 4x + 4)$$
$$= 2(2x^2 + 4x + 3)$$
$$= 4x^2 + 8x + 6.$$

Chapter 7: Practices

Simplify each polynomial.

1) $3(6x+4) =$
2) $5(3x-8) =$
3) $x(7x+2)+9x =$
4) $6x(x+3)+5x =$
5) $6x(3x+1)-5x =$
6) $x(3x-4)+3x^2-6 =$
7) $x^2-5-3x(x+8) =$
8) $2x^2+7-6x(2x+5) =$

Add or subtract polynomials.

9) $(x^2+3)+(2x^2-4) =$
10) $(3x^2-6x)-(x^2+8x) =$
11) $(4x^3-3x^2)+(2x^3-5x^2) =$
12) $(6x^3-7x)-(5x^3-3x) =$
13) $(10x^3+4x^2)+(14x^2-8) =$
14) $(4x^3-9)-(3x^3-7x^2) =$
15) $(9x^3+3x)-(6x^3-4x) =$
16) $(7x^3-5x)-(3x^3+5x) =$

Use algebra tiles to simplify polynomials.

17) $(2x^2-3x+3)-(x^2-x-1)$
18) $(2x^2+2x+5)+(x^2+2x+1)$

Find the products.

19) $3x^2 \times 8x^3 =$
20) $2x^4 \times 9x^3 =$
21) $-4a^4b \times 2ab^3 =$
22) $(-7x^3yz) \times (3xy^2z^4) =$
23) $-2a^5bc \times 6a^2b^4 =$
24) $9u^3t^2 \times (-2ut) =$
25) $12x^2z \times 3xy^3 =$
26) $11x^3z \times 5xy^5 =$

27) $-6a^3bc \times 5a^4b^3 =$

28) $-4x^6y^2 \times (-12xy) =$

Simplify each expression.

29) $(7x^2y^3)(3x^4y^2) =$

30) $(6x^3y^2)(4x^4y^3) =$

31) $(10x^8y^5)(3x^5y^7) =$

32) $(15a^3b^2)(2a^3b^8) =$

33) $\frac{42x^4y^2}{6x^3y} =$

34) $\frac{49x^5y^6}{7x^2y} =$

35) $\frac{63x^{15}y^{10}}{9x^8y^6} =$

36) $\frac{35x^8y^{12}}{5x^4y^8} =$

Find each product.

37) $3x(5x - y) =$

38) $2x(4x + y) =$

39) $7x(x - 3y) =$

40) $x(2x^2 + 2x - 4) =$

41) $5x(3x^2 + 8x + 2) =$

42) $7x(2x^2 - 9x - 5) =$

Use the area model to find each product.

43) $3x(x + 2)$

44) $(a - 3)(2a + 2)$

Find each product.

45) $(x - 3)(x + 3) =$

46) $(x - 6)(x + 6) =$

47) $(x + 10)(x + 4) =$

48) $(x - 6)(x + 7) =$

49) $(x + 2)(x - 5) =$

50) $(x - 10)(x + 3) =$

Use algebra tiles to simplify.

51) $(x+1)(x+6)$

52) $(2x+1)(x-4)$

Factor each trinomial.

53) $x^2+6x+8=$

54) $x^2+3x-10=$

55) $x^2+2x-48=$

56) $x^2-10x+16=$

57) $2x^2-10x+12=$

58) $3x^2-10x+3=$

Factor each expression.

59) $4x^2-4x-8$

60) $6x^2+37x+6$

61) $16x^2+60x-100$

62) $4x^2-17x+4$

Use a graph to factor the following polynomial.

63) x^2-4

64) $-(x+2)^2$

Factor each completely.

65) $36x^2-121$

66) $-36x^4+4x^2$

67) $-36x^2+400=$

68) $49x^2-56x+16=$

69) $1-x^2=$

70) $81x^4-900x^2=$

Find the perimeter.

71)

$3x+6$

$(x+2)$

72)

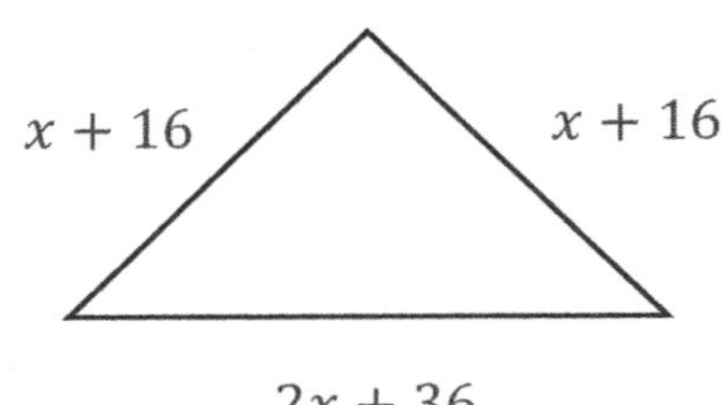

Chapter 7: Answers

1) $18x + 12$
2) $15x - 40$
3) $7x^2 + 11x$
4) $6x^2 + 23x$
5) $18x^2 + x$
6) $6x^2 - 4x - 6$
7) $-2x^2 - 24x - 5$
8) $-10x^2 - 30x + 7$
9) $3x^2 - 1$
10) $2x^2 - 14x$
11) $6x^3 - 8x^2$
12) $x^3 - 4x$
13) $10x^3 + 18x^2 - 8$
14) $x^3 + 7x^2 - 9$
15) $3x^3 + 7x$
16) $4x^3 - 10x$

17) $x^2 - 2x + 4$

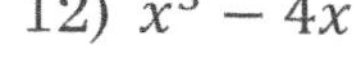

18) $3x^2 + 4x + 6$

19) $24x^5$
20) $18x^7$
21) $-8a^5b^4$
22) $-21x^4y^3z^5$
23) $-12a^7b^5c$
24) $-18u^4t^3$
25) $36x^3y^3z$
26) $55x^4y^5z$
27) $-30a^7b^4c$
28) $48x^7y^3$
29) $21x^6y^5$
30) $24x^7y^5$

31) $30x^{13}y^{12}$
32) $30a^6b^{10}$
33) $7xy$
34) $7x^3y^5$
35) $7x^7y^4$
36) $7x^4y^4$
37) $15x^2 - 3xy$
38) $8x^2 + 2xy$
39) $7x^2 - 21xy$
40) $2x^3 + 2x^2 - 4x$
41) $15x^3 + 40x^2 + 10x$
42) $14x^3 - 63x^2 - 35x$

43) $3x^2 + 6x$

	$3x$
x	$3x^2$
2	$6x$

44) $2a^2 - 4a - 6$

	$2a$	2
a	$2a^2$	$2a$
-3	$-6a$	-6

45) $x^2 - 9$
46) $x^2 - 36$

Effortless Math Education

47) $x^2 + 14x + 40$

48) $x^2 + x - 42$

49) $x^2 - 3x - 10$

50) $x^2 - 7x - 30$

51) $x^2 + 7x + 6$

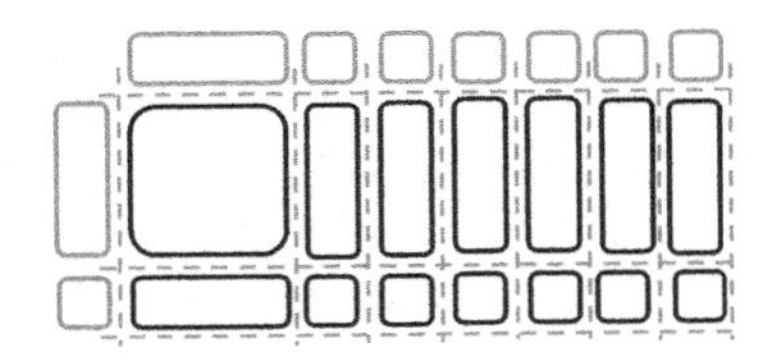

52) $2x^2 - 7x - 4$

53) $(x + 4)(x + 2)$

54) $(x + 5)(x - 2)$

55) $(x - 6)(x + 8)$

56) $(x - 8)(x - 2)$

57) $(2x - 4)(x - 3)$

58) $(3x - 1)(x - 3)$

59) $4(x + 1)(x - 2)$

60) $(x + 6)(6x + 1)$

61) $4(x + 5)(4x - 5)$

62) $(x - 4)(4x - 1)$

63) $x = \pm 2$

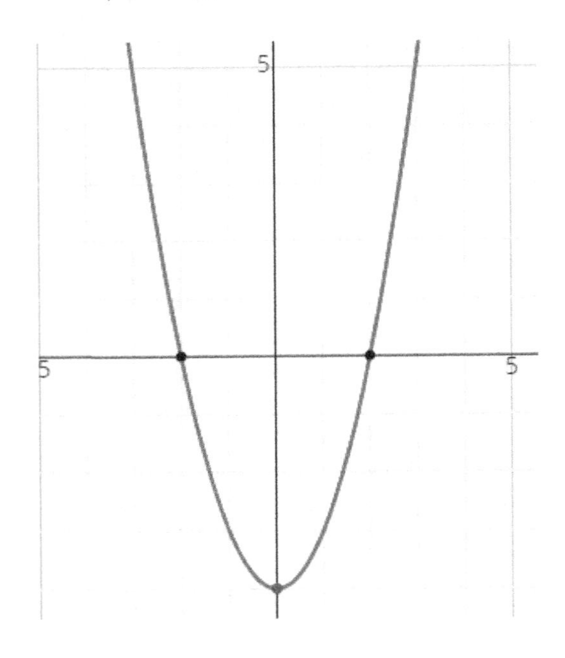

64) 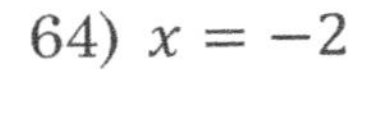

$x = -2$

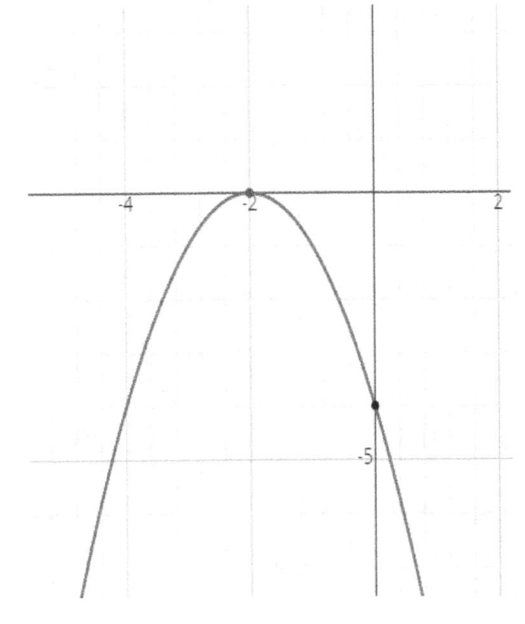

65) $(6x - 11)(6x + 11)$

66) $4x^2(1 - 3x)(1 + 3x)$

67) $4(10 + 3x)(10 - 3x)$

68) $(7x - 4)^2$

69) $(1 + x)(1 - x)$

70) $9x^2(3x + 10)(3x - 10)$

71) $8x + 16$

72) $4x + 68$

CHAPTER

8 Relations and Functions

Math topics that you'll learn in this chapter:

- ☑ Function Notation and Evaluation
- ☑ Adding and Subtracting Functions
- ☑ Multiplying and Dividing Functions
- ☑ Composition of Functions
- ☑ Evaluate an Exponential Function
- ☑ Match Exponential Functions and Graphs
- ☑ Write Exponential Functions: Word Problems
- ☑ Function Inverses
- ☑ Domain and Range of Relations
- ☑ Rate of Change and Slope
- ☑ Complete a Function Table from an Equation

Function Notation and Evaluation

- Functions are mathematical operations that assign unique outputs to given inputs.
- Function notation is the way a function is written. It is meant to be a precise way of giving information about the function without a rather lengthy written explanation.
- The most popular function notation is $f(x)$ which is read "f of x". Any letter can name a function. For example: $g(x)$, $h(x)$, etc.
- To evaluate a function, plug in the input (the given value or expression) for the function's variable (place holder, x).

Examples:

Example 1. Evaluate: $f(x) = x + 6$, find $f(2)$

Solution: Substitute x with 2:

Then: $f(x) = x + 6 \rightarrow f(2) = 2 + 6 \rightarrow f(2) = 8$.

Example 2. Evaluate: $w(x) = 3x - 1$, find $w(4)$.

Solution: Substitute x with 4:

Then: $w(x) = 3x - 1 \rightarrow w(4) = 3(4) - 1 = 12 - 1 = 11$.

Example 3. Evaluate: $f(x) = 2x^2 + 4$, find $f(-1)$.

Solution: Substitute x with -1:

Then: $f(x) = 2x^2 + 4 \rightarrow f(-1) = 2(-1)^2 + 4 \rightarrow f(-1) = 2 + 4 = 6$.

Example 4. Evaluate: $h(x) = 4x^2 - 9$, find $h(2a)$.

Solution: Substitute x with $2a$:

Then: $h(x) = 4x^2 - 9 \rightarrow h(2a) = 4(2a)^2 - 9 \rightarrow h(2a) = 4(4a^2) - 9 = 16a^2 - 9$.

Adding and Subtracting Functions

- Just like we can add and subtract numbers and expressions, we can add or subtract functions and simplify or evaluate them. The result is a new function.
- For two functions $f(x)$ and $g(x)$, we can create two new functions:

$$(f+g)(x) = f(x) + g(x) \text{ and } (f-g)(x) = f(x) - g(x)$$

Examples:

Example 1. $g(x) = 2x - 2$, $f(x) = x + 1$, find: $(g + f)(x)$.

Solution: $(g + f)(x) = g(x) + f(x)$

Then: $(g + f)(x) = (2x - 2) + (x + 1) = 2x - 2 + x + 1 = 3x - 1$.

Example 2. $f(x) = 4x - 3$, $g(x) = 2x - 4$, find: $(f - g)(x)$.

Solution: $(f - g)(x) = f(x) - g(x)$

Then: $(f - g)(x) = (4x - 3) - (2x - 4) = 4x - 3 - 2x + 4 = 2x + 1$.

Example 3. $g(x) = x^2 + 2$, $f(x) = x + 5$, find: $(g + f)(x)$.

Solution: $(g + f)(x) = g(x) + f(x)$

Then: $(g + f)(x) = (x^2 + 2) + (x + 5) = x^2 + x + 7$.

Example 4. $f(x) = 5x^2 - 3$, $g(x) = 3x + 6$, find: $(f - g)(3)$.

Solution: $(f - g)(x) = f(x) - g(x)$

Then: $(f - g)(x) = (5x^2 - 3) - (3x + 6) = 5x^2 - 3 - 3x - 6 = 5x^2 - 3x - 9$.

Substitute x with 3: $(f - g)(3) = 5(3)^2 - 3(3) - 9 = 45 - 9 - 9 = 27$.

Example 5. $g(x) = x^2 - 4$, $f(x) = 2x + 3$, find: $(g + f)(x)$.

Solution: $(g + f)(x) = g(x) + f(x)$

Then: $(g + f)(x) = (x^2 - 4) + (2x + 3) = x^2 - 4 + 2x + 3 = x^2 + 2x - 1$.

Multiplying and Dividing Functions

- Just like we can multiply and divide numbers and expressions, we can multiply and divide two functions and simplify or evaluate them.
- For two functions $f(x)$ and $g(x)$, we can create two new functions:

$$(f.g)(x) = f(x).g(x) \text{ and } \left(\frac{f}{g}\right)(x) = \frac{f(x)}{g(x)}$$

Examples:

Example 1. $g(x) = x + 3$, $f(x) = x + 4$, find: $(g.f)(x)$.

Solution:

$(g.f)(x) = g(x).f(x) = (x + 3)(x + 4) = x^2 + 4x + 3x + 12 = x^2 + 7x + 12.$

Example 2. $f(x) = x + 6$, $h(x) = x - 9$, find: $\left(\frac{f}{h}\right)(x)$.

Solution: $\left(\frac{f}{h}\right)(x) = \frac{f(x)}{h(x)} = \frac{x+6}{x-9}$.

Example 3. $g(x) = x + 7$, $f(x) = x - 3$, find: $(g.f)(2)$.

Solution: $(g.f)(x) = g(x).f(x) = (x + 7)(x - 3) = x^2 - 3x + 7x - 21.$

Then: $g(x).f(x) = x^2 + 4x - 21.$

Substitute x with 2: $(g.f)(2) = (2)^2 + 4(2) - 21 = 4 + 8 - 21 = -9.$

Example 4. $f(x) = x + 3$, $h(x) = 2x - 4$, find: $\left(\frac{f}{h}\right)(3)$.

Solution: $\left(\frac{f}{h}\right)(x) = \frac{f(x)}{h(x)} = \frac{x+3}{2x-4}$.

Substitute x with 3: $\left(\frac{f}{h}\right)(3) = \frac{3+3}{2(3)-4} = \frac{6}{2} = 3.$

Example 5. $g(x) = x + 5$, $f(x) = x - 2$, find: $(g.f)(4)$.

Solution: $(g.f)(x) = g(x).f(x) = (x + 5)(x - 2) = x^2 + 3x - 10.$

Substitute x with 4: $(g.f)(4) = (4)^2 + 3(4) - 10 = 16 + 12 - 10 = 18.$

Composition of Functions

- "Composition of functions" simply means combining two or more functions in a way where the output from one function becomes the input for the next function.

- The notation used for a composition is: $(fog)(x) = f(g(x))$ and is read

  "f composed with g of x" or "f of g of x".

Examples:

Example 1. Using $f(x) = 2x + 3$ and $g(x) = 5x$, find: $(fog)(x)$.

Solution: $(fog)(x) = f(g(x))$. Then:

$$(fog)(x) = f(g(x)) = f(5x).$$

Now find $f(5x)$ by substituting x with $5x$ in $f(x)$ function.

Then:

$$f(x) = 2x + 3; (x \rightarrow 5x) \rightarrow f(5x) = 2(5x) + 3 = 10x + 3.$$

Example 2. Using $f(x) = 3x - 1$ and $g(x) = 2x - 2$, find: $(gof)(5)$.

Solution: $(fog)(x) = f(g(x))$. Then:

$$(gof)(x) = g(f(x)) = g(3x - 1),$$

Now substitute x in $g(x)$ *by* $(3x - 1)$.

Then:

$$g(3x - 1) = 2(3x - 1) - 2 = 6x - 2 - 2 = 6x - 4.$$

Substitute x with 5: $(gof)(5) = g(f(5)) = 6(5) - 4 = 30 - 4 = 26.$

Example 3. Using $f(x) = 2x^2 - 5$ and $g(x) = x + 3$, find: $f(g(3))$.

Solution: First, find $g(3)$:

$$g(x) = x + 3 \rightarrow g(3) = 3 + 3 = 6.$$

Then: $f(g(3)) = f(6)$.

Now, find $f(6)$ by substituting x with 6 in $f(x)$ function.

$$f(g(3)) = f(6) = 2(6)^2 - 5 = 2(36) - 5 = 67.$$

Evaluate an Exponential Function

- For any real number x, an exponential function is an equation with the form $f(x) = ab^x$, where a is a non-zero real number and b is a positive real number.
- To evaluate the value of an exponential function, it's enough to substitute the given input for the independent variable.

Examples:

Example 1. Let $f(x) = 3^x$. What is $f(2)$?

Solution: To solve, it is enough to substitute $x = 2$, in the equation $f(x) = 3^x$. So, $f(2) = 3^2 = 9$.

Example 2. Let $f(x) = 7(2)^x$. Evaluate $f(3)$.

Solution: To evaluate $f(3)$, plug 3 in equation $f(x) = 7(2)^x$ instead of the independent variable x. Therefore,

$$f(3) = 7(2)^3$$
$$= 7(8) = 56.$$

Example 3. Use the following function to find $f(6)$.

$$f(x) = -2(3)^{\frac{x}{2}-1}$$

Solution: First, substitute $x = 6$ in the equation,

$$f(6) = -2(3)^{\frac{6}{2}-1} = -2(3)^{3-1} = -2(3)^2$$
$$= -2(9) = -18.$$

Example 4. Use the following function to find $f(2)$.

$$f(x) = 9\left(\frac{1}{3}\right)^{2x-1} + 1$$

Solution: To solve, plug 2 into $f(x) = 9\left(\frac{1}{3}\right)^{2x-1} + 1$ instead of x. So,

$$f(2) = 9\left(\frac{1}{3}\right)^{2(2)-1} + 1 = 9\left(\frac{1}{3}\right)^{2(2)-1} + 1 = 9\left(\frac{1}{3}\right)^{4-1} + 1 = 9\left(\frac{1}{3}\right)^3 + 1$$
$$= 9\left(\frac{1}{27}\right) + 1 = \frac{1}{3} + 1 = \frac{4}{3}.$$

Match Exponential Functions and Graphs

- To match an exponential function $y = b^x$ with its graph and vice versa, use points to identify the important parameters of the graph of an exponential function, which are:
- Look for the relationship as growth or decay. By checking:
  - If $b > 1$, the function is growing.
  - If $0 < b < 1$, the function is decaying.
- Evaluate the value of the function at a few inputs to match some points on the graph, like the y −intercept.
- Determine the end behavior of the function.

Example:

Match each exponential function to its graph.

$$f(x) = \left(\frac{1}{2}\right)^x,\ g(x) = \left(\frac{3}{2}\right)^x,\ h(x) = \left(\frac{1}{3}\right)^x$$

A

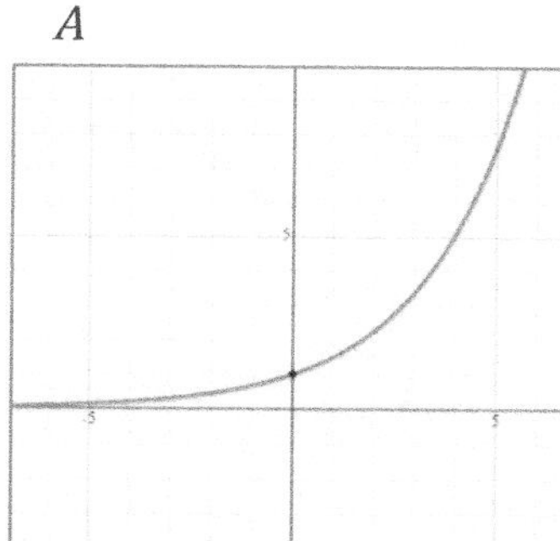

B

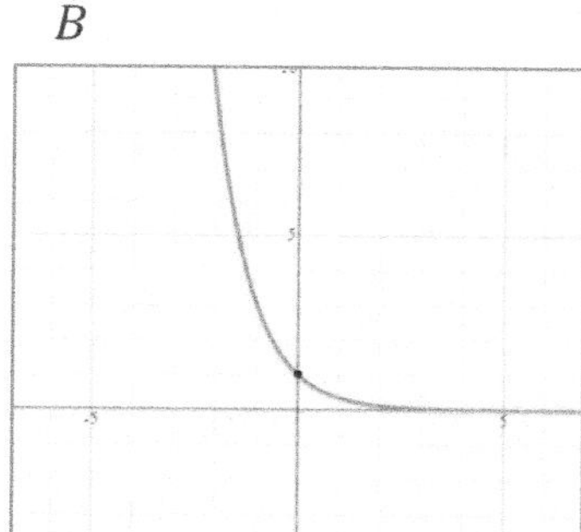

C

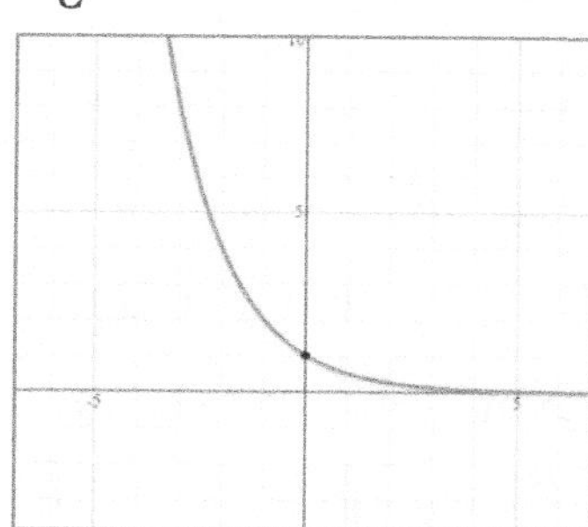

Solution: According to the base value of the exponential equation, notice that one of the functions is growing and the other two functions are decaying. So, the function $g(x) = \left(\frac{3}{2}\right)^x$ can be equivalent to graph A. Now, evaluate the two remaining functions at a few inputs to find some points on the graph. Choose a point such as -1 to plug into the equation. Therefore, start with $f(x) = \left(\frac{1}{2}\right)^x$.

$$f(-1) = \left(\frac{1}{2}\right)^{-1} = 2.$$

You can see that the ordered pair $(-1,2)$ for the function $f(x) = \left(\frac{1}{2}\right)^x$ is equivalent to a point on graph C. In the same way, substitute -1 in $h(x) = \left(\frac{1}{3}\right)^x$. Therefore, $h(-1) = \left(\frac{1}{3}\right)^{-1} = 3$. That is, graph B represented the function $h(x)$.

bit.ly/3ZkcfK

Write Exponential Functions: Word Problems

To solve the word problems corresponding to the exponential function, follow the steps which are:

Step 1: Check that it changes at the constant ratio.
Step 2: Identify the given values such as the ratio and the initial amount.
Step 3: Substitute in the exponential formula.
Step 4: Evaluate the requested values.

Examples:

Example 1. In a laboratory sample, if it starts with 100 bacteria which can double every hour, how many bacteria will there be after 6 hours?

Solution: Since it grows at the constant ratio of 2, you have to use the formula of exponential to find the number of bacteria in this sample. Use the formula,

$$y = ab^n$$

Where $a = 100$ is the initial number of bacteria, and b is equal to 2. Substitute the given value in the formula, then:

$$y = 100(2)^n$$

Now, plug $n = 6$ into the equation. Therefore, $y = 100(2)^6 = 100(64) = 6{,}400$.

Example 2. Mrs. Shelby's new car cost $12,000. It depreciates in value by about 10% each year. Write an equation that would indicate the value of the car at the tth year. How much will her car be worth in 8 years?

Solution: To write the equation of the value of the car at the tth year, use the formula given as follow: $y = ab^x$, where a is the initial amount, and b is the ratio of changes. The initial cost of the car is $12,000. So, the equation is

$$y = 12{,}000(0.9)^t$$

Now, plug $t = 8$ in $y = 12{,}000(0.9)^t$. Therefore,

$$y = 12{,}000(0.9)^8$$

$$\cong 12{,}000(0.43) = 5{,}160.$$

Function Inverses

- An inverse function is a function that reverses another function: if the function f applied to an input x gives a result of y, then applying its inverse function g to y gives the result x.
- $f(x) = y$ if and only if $g(y) = x$.
- The inverse function of $f(x)$ is usually shown by $f^{-1}(x)$.

Examples:

Example 1. Find the inverse of the function: $f(x) = 2x - 1$.

Solution: First, replace $f(x)$ with y: $y = 2x - 1$. Then, replace all x's with y and all y's with x: $x = 2y - 1$.

Now, solve for y: $x = 2y - 1 \rightarrow x + 1 = 2y \rightarrow \frac{1}{2}x + \frac{1}{2} = y$.

Finally replace y with $f^{-1}(x)$:

$$f^{-1}(x) = \frac{1}{2}x + \frac{1}{2}.$$

Example 2. Find the inverse of the function: $g(x) = \frac{1}{5}x + 3$.

Solution: $g(x) = \frac{1}{5}x + 3 \rightarrow y = \frac{1}{5}x + 3$,

replace all x's with y and all y's with x; $x = \frac{1}{5}y + 3$,

solve for y:

$$x - 3 = \frac{1}{5}y \rightarrow 5(x - 3) = y \rightarrow g^{-1}(x) = 5x - 15.$$

Example 3. Find the inverse of the function: $h(x) = \sqrt{x} + 6$.

Solution: $h(x) = \sqrt{x} + 6 \rightarrow y = \sqrt{x} + 6$,

replace all x's with y and all y's with x;

$$x = \sqrt{y} + 6 \rightarrow x - 6 = \sqrt{y} \rightarrow (x - 6)^2 = (\sqrt{y})^2 \rightarrow x^2 - 12x + 36 = y.$$

Then:

$h^{-1}(x) = x^2 - 12x + 36$.

Domain and Range of Relations

- A relation is defined as a set or the desired set's connection. An ordered pair commonly is named a point, and a relation is a set of these ordered pairs. In fact, input and output values in a relationship are shown in ordered pairs. In other words, a relation is a kind of rule that connects a component or value from one set to a component or value from the other set.
- The domain refers to all of the input or independent values that are put into a relation or a function. The range refers to all of the output or dependent values that leave from a relation or a function.
- A function is a group of ordered pairs that each input element just is related to an output element. In other words, in a function 2 input values can be connected to the same output value but it's not possible that 2 output values are connected to the same input value.
- When you have a graph, the x −coordinates of the graph are domain and the y −coordinates of the graph are range. The x −coordinates are the domain value, when you put them into the function, the output values you have found will be placed on the y −axis.

Examples:

Example 1. What is the domain and range of the following relation: $\{(7,2),(-3,4),(4,-1),(5,3),(8,5)\}$?

Solution: The domain contains x −values of a relation and the range includes y −values of a relationship. So, Domain $=\{7,-3,4,5,8\}$ and Range $=\{2,4,-1,3,5\}$.

Example 2. Find the domain and range of the following relation: $R=\{(4x+1,x-\ 2):x\in\{-2,-1,0,2,3\}$

Solution: Given the relation $R=\{4x+1,x-2\}$ where x belongs to the set $\{-2,-1,0,2,3\}$, let's determine the output values for each value of x. For $x=-2: 4(-2)+1=-7,(-2)-2=-4$. For $x=-1: 4(-1)+1=-3,(-1)-2=-3$. For $x=0: 4(0)+1=1, 0-2=-2$. For $x=2: 4(2)+1=9, 2-2=0$. For $x=3: 4(3)+1=13, 3-2=1$.

From the calculations we have: $R=\{(-7,-4),(-3,-3),(1,-2),(9,0),(13,1)\}$

Domain $=\{-7,-3,1,9,13\}$.

Range $=\{-4,-3,-2,0,1\}$,

Rate of Change and Slope

- A ratio that uses to compare the change in y −values to the change in the x −values is called the rate of change. y −values are considered as the dependent variables and x − values are considered as the independent variables. The rate of change is the line's slope when the rate of change is constant and linear. The line's slope can be negative, positive, zero, or undefined.

- The direction of a line's slope can also describe a slope of a line. If a line is a rising line from left to right its slope is positive. If a line is a falling line from left to right, its slope is negative. In the case when you have a horizontal line it means that no change happens, so the slope of the line is zero. when you have a vertical line, this relationship is not a function, and the slope is undefined. It's because there are many y −values for one x −value.

- The variable m is the sign for the slope of a line. It is stated by the ratio of the subtraction of y −variables to the subtraction of x −variables. It means if $(x_1, y_1)(x_2, y_2)$ are the coordinates of 2 points on a line, then $m = \frac{y_2 - y_1}{x_2 - x_1}$.

Example:

The following table shows the number of cars sold by a company in different years. Find the rate of change in car sales for each time interval. Determine which time interval has the greatest rate.

Year	2005	2010	2015	2020	2022
Number of sold cars	35	45	47	67	85

Solution: First, find the dependent and independent variables: years are independent variables and the number of sold cars are dependent variables. Now find the rate of changes: 2005 to 2010 $\rightarrow \frac{change\ in\ the\ number\ of\ sold\ cars}{change\ in\ years} = \frac{45-35}{2010-2005} = 2$. 2010 to 2015 $\rightarrow \frac{change\ in\ the\ number\ of\ sold\ cars}{change\ in\ years} = \frac{47-45}{2015-2010} = 0.4$. 2015 to 2020 $\rightarrow \frac{change\ in\ the\ number\ of\ sold\ cars}{change\ in\ years} = \frac{67-47}{2020-2015} = 4$. 2022 to 2020 $\rightarrow \frac{change\ in\ the\ number\ of\ sold\ cars}{change\ in\ years} = \frac{85-67}{2022-2020} = 9$. Car sales have been at their greatest rate from 2022 to 2020. The slope of the line is positive because as the time period increases, the amount of car sales also increases.

Complete a Function Table from an Equation

- One of the main ways to show a relationship in mathematics is by using a function table.
- To complete a function table of the given equation:

Step 1: Consider the input and output in the table.

Step 2: Substitute the given value for the input.

Step 3: Evaluate the output of the equation.

Example:

Complete the table.

$f(x) = 2x - 1$	
x	$f(x)$
-2	
0	
1	

Solution: Look at the function table. Clearly, the first value of the input in the table is -2. Evaluate $f(x) = 2x - 1$ for $x = -2$.

$$f(-2) = 2(-2) - 1$$

$$= -4 - 1 = -5.$$

$f(x) = 2x - 1$	
x	$f(x)$
-2	-5
0	
1	

When $x = -2$, then $f(-2) = -5$. Complete the first row of the table.

Similarly, evaluate $f(x) = 2x - 1$ for $x = 0$ and $x = 1$, respectively. So,

$$f(0) = 2(0) - 1 = 0 - 1 = -1,$$

$$f(1) = 2(1) - 1 = 2 - 1 = 1.$$

Enter the obtained values in the table.

$f(x) = 2x - 1$	
x	$f(x)$
-2	-5
0	-1
1	1

Chapter 8: Practices

Evaluate each function.

1) $f(x) = x - 2$, find $f(-1)$
2) $g(x) = 2x + 4$, find $g(3)$
3) $g(n) = 2n - 8$, find $g(-1)$
4) $h(n) = n^2 - 1$, find $h(-2)$
5) $f(x) = x^2 + 12$, find $f(5)$
6) $g(x) = 2x^2 - 9$, find $g(-2)$
7) $w(x) = 2x^2 - 4x$, find $w(2n)$
8) $p(x) = 4x^3 - 10$, find $p(-3a)$

Perform the indicated operation.

9) $g(x) = x - 2$
   $h(x) = 2x + 6$
   Find: $(h + g)(3)$

   ______

10) $f(x) = 3x + 2$
    $g(x) = -x - 6$
    Find: $(f + g)(2)$

    ______

11) $f(x) = 5x + 8$
    $g(x) = 3x - 12$
    Find: $(f - g)(-2)$

    ______

12) $h(x) = 2x^2 - 10$
    $g(x) = 3x + 12$
    Find: $(h + g)(3)$

    ______

13) $g(x) = 12x - 8$
    $h(x) = 3x^2 + 14$
    Find: $(h - g)(x)$

    ______

14) $h(x) = -2x^2 - 18$
    $g(x) = 4x^2 + 15$
    Find: $(h - g)(a)$

    ______

Perform the indicated operation.

15) $g(x) = x - 5$
    $h(x) = x + 6$
    Find: $(g.h)(-1)$

    ____

16) $f(x) = 2x + 2$
    $g(x) = -x - 6$
    Find: $(\frac{f}{g})(-2)$

    ____

17) $f(x) = 5x + 3$
    $g(x) = 2x - 4$
    Find: $(\frac{f}{g})(5)$

    ____

18) $h(x) = x^2 - 2$
    $g(x) = x + 4$
    Find: $(g.h)(3)$

    ____

19) $g(x) = 4x - 12$
    $h(x) = x^2 + 4$
    Find: $(g.h)(-2)$

    ____

20) $h(x) = 3x^2 - 8$
    $g(x) = 4x + 6$
    Find: $(\frac{h}{g})(-4)$

    ____

Solve.

21) $f(x) = 2x$

$g(x) = x + 3$

Find: $(fog)(2)$

_____

22) $f(x) = x + 2$

$g(x) = x - 6$

Find: $(fog)(-1)$

_____

23) $f(x) = 3x$

$g(x) = x + 4$

Find: $(gof)(4)$

_____

24) $h(x) = 2x - 2$

$g(x) = x + 4$

Find: $(goh)(2)$

_____

25) $f(x) = 2x - 8$

$g(x) = x + 10$

Find: $(fog)(-2)$

_____

26) $f(x) = x^2 - 8$

$g(x) = 2x + 3$

Find: $(gof)(4)$

_____

Use the following function to find: $f(x) = 3x(\frac{1}{2})^{2x+2}$

27) $f(2)$

28) $f(4)$

Match each exponential function to its graph.

29) $f(x) = -4(3)^x$, $f(x) = 2^x + 5$, $f(x) = 3(2)^x + 3$

A

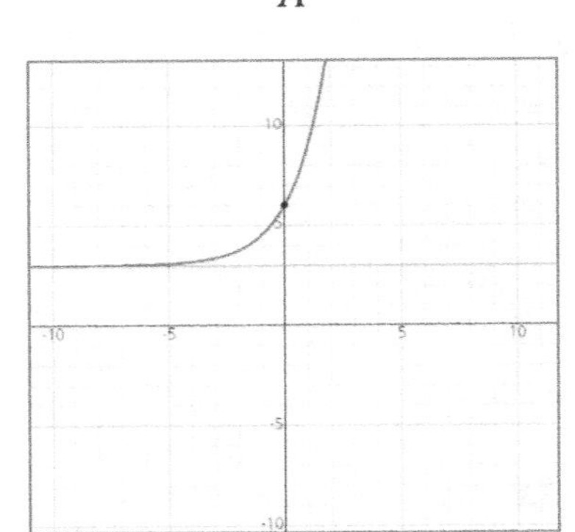

B

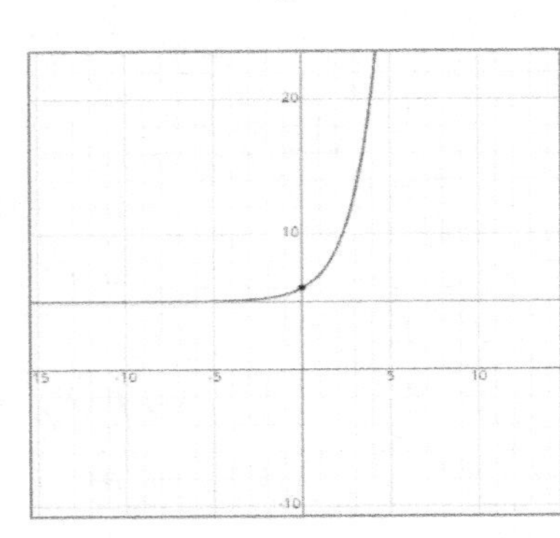

C

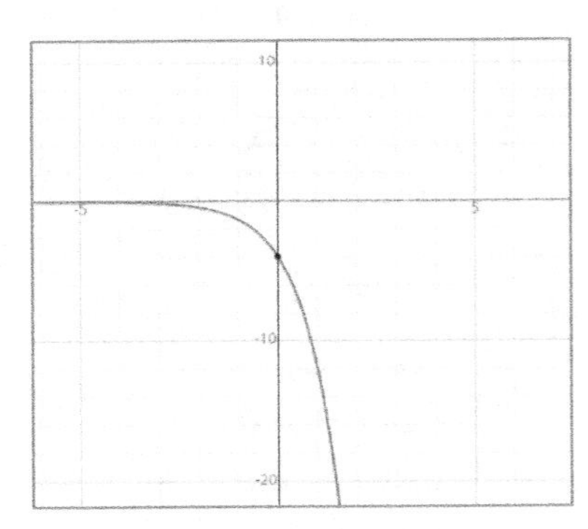

Solve.

30) As of 2019, the world population is 8.716 billion and growing at a rate of 1.2% per year. Write an equation to model population growth, where $p(t)$ is the population in billions of people and t is the time in years.

31) You decide to buy a used car that costs $20,000. You have heard that the car may depreciate at a rate of 10% per year. At this rate, how much will the car be worth in 6 years?

Find the inverse of each function.

32) $f(x) = -\frac{1}{x} - 9$

$f^{-1}(x) =$ ________

33) $g(x) = \sqrt{x} - 2$

$g^{-1}(x) =$ ________

34) $h(x) = -\frac{5}{x+3}$

$h^{-1}(x) =$ ________

35) $f(x) = 6x + 6$

$f^{-1}(x) =$ ________

Find the domain and range of each relation.

36) $\{(1,-1),(2,-4),(0,5),(-1,6)\}$

37) $\{(10,-5),(-16,-8),(-4,19),(16,7),(6,-14)\}$

38) $\{(4,7),(-15,6),(-20,9),(13,8),(7,5)$

Solve.

39) Average food preparation time in a restaurant was tracked daily as part of an efficiency improvement program.

Day	Food preparation time (minutes)
Tuesday	45
Wednesday	49
Thursday	32
Friday	15
Saturday	25

According to the table, what was the rate of change between Tuesday and Wednesday?

Complete the table.

40)

$f(x) = 3x - 2$	
x	$f(x)$
-3	
0	
2	

41)

$f(x) = 2x$	
x	$f(x)$
1	
2	
3	

Chapter 8: Answers

1) -3
2) 10
3) -10
4) 3
5) 37
6) -1
7) $8n^2 - 8n$
8) $-108a^3 - 10$
9) 13
10) 0
11) 16
12) 29
13) $3x^2 - 12x + 22$
14) $-6a^2 - 33$
15) -30
16) $\frac{1}{2}$
17) $\frac{14}{3}$
18) 49
19) -160
20) -4
21) 10
22) -5
23) 16
24) 6
25) 8
26) 19
27) $\frac{3}{32}$
28) $\frac{3}{256}$
29) $A = 3(2)^x + 3$,

    $B = 2^x + 5$,

    $C = -4(3)^x$
30) $p(t) = 8.716(1 + 0.012)^t$
31) $A = 20{,}000(1 - 0.1)^6$
32) $-\frac{1}{x+9}$
33) $x^2 + 4x + 4$
34) $-\frac{5}{x} - 3$
35) $\frac{x-6}{6}$

36) $D = (1,2,0,-1)$,

$R = (-1,-4,5,6)$

37) $D = (10,-16,-4,16,6)$,

$R = (-5,-8,19,7,-14)$

38) $D = (4,-15,-20,13,7)$,

$R = (7,6,9,8,5)$

39) 4

40)

$f(x) = 3x - 2$	
x	$f(x)$
-3	-11
0	-2
2	4

41).

$f(x) = 2x$	
x	$f(x)$
1	2
2	4
3	6

CHAPTER

9 Radical Expressions

Math topics that you'll learn in this chapter:

- ☑ Simplifying Radical Expressions
- ☑ Adding and Subtracting Radical Expressions
- ☑ Multiplying Radical Expressions
- ☑ Rationalizing Radical Expressions
- ☑ Radical Equations
- ☑ Domain and Range of Radical Functions
- ☑ Simplify Radicals with Fractions

Simplifying Radical Expressions

- Find the prime factors of the numbers or expressions inside the radical.
- Use radical properties to simplify the radical expression:

$$\sqrt[n]{x^a} = x^{\frac{a}{n}},\ \sqrt[n]{xy} = x^{\frac{1}{n}} \times y^{\frac{1}{n}},\ \sqrt[n]{\frac{x}{y}} = \frac{x^{\frac{1}{n}}}{y^{\frac{1}{n}}},\text{ and } \sqrt[n]{x} \times \sqrt[n]{y} = \sqrt[n]{xy}$$

Examples:

Example 1. Find the square root of $\sqrt{144x^2}$.

Solution: Find the factor of the expression $144x^2$: $144 = 12 \times 12$ and $x^2 = x \times x$, now use radical rule: $\sqrt[n]{a^n} = a$.

Then: $\sqrt{12^2} = 12$ and $\sqrt{x^2} = x$.

Finally:

$$\sqrt{144x^2} = \sqrt{12^2} \times \sqrt{x^2} = 12 \times x = 12x.$$

Example 2. Write this radical in exponential form. $\sqrt[3]{x^4}$

Solution: To write a radical in exponential form, use this rule: $\sqrt[n]{x^a} = x^{\frac{a}{n}}$.

Then:

$$\sqrt[3]{x^4} = x^{\frac{4}{3}}.$$

Example 3. Simplify. $\sqrt{8x^3}$

Solution: First factor the expression $8x^3$: $8x^3 = 2^3 \times x \times x \times x$, we need to find perfect squares: $8x^3 = 2^2 \times 2 \times x^2 \times x = 2^2 \times x^2 \times 2x$,

Then:

$$\sqrt{8x^3} = \sqrt{2^2 \times x^2} \times \sqrt{2x}.$$

Now use radical rule: $\sqrt[n]{a^n} = a$.

Then:

$$\sqrt{2^2 \times x^2} \times \sqrt{(2x)} = 2x \times \sqrt{2x} = 2x\sqrt{2x}.$$

Adding and Subtracting Radical Expressions

- Only numbers and expressions that have the same radical part can be added or subtracted.
- Remember, combining "unlike" radical terms is not possible.
- For numbers with the same radical part, just add or subtract factors outside the radicals.

Examples:

Example 1. Simplify: $6\sqrt{2}+5\sqrt{2}$.

Solution: Since we have the same radical parts, then we can add these two radicals. Add like terms:

$$6\sqrt{2}+5\sqrt{2}=11\sqrt{2}.$$

Example 2. Simplify: $2\sqrt{8}-2\sqrt{2}$.

Solution: The two radical parts are not the same. First, we need to simplify the $2\sqrt{8}$. Then:

$$2\sqrt{8}=2\sqrt{4\times 2}=2(\sqrt{4})(\sqrt{2})=4\sqrt{2}.$$

Now, combine like terms:

$$2\sqrt{8}-2\sqrt{2}=4\sqrt{2}-2\sqrt{2}=2\sqrt{2}.$$

Example 3. Simplify: $8\sqrt{27}+5\sqrt{3}$.

Solution: The two radical parts are not the same. First, we need to simplify the $8\sqrt{27}$. Then:

$$8\sqrt{27}=8\sqrt{9\times 3}=8(\sqrt{9})(\sqrt{3})=24\sqrt{3}.$$

Now, add:

$$8\sqrt{27}+5\sqrt{3}=24\sqrt{3}+5\sqrt{3}=29\sqrt{3}.$$

bit.ly/2PkJfTA
Find more at

Multiplying Radical Expressions

To multiply radical expressions:

- Multiply the numbers and expressions outside of the radicals.
- Multiply the numbers and expressions inside the radicals.
- Simplify if needed.

Examples:

Example 1. Evaluate. $2\sqrt{5} \times \sqrt{3}$

Solution: Multiply the numbers outside of the radicals and the radical parts. Then:

$$2\sqrt{5} \times \sqrt{3} = (2 \times 1) \times (\sqrt{5} \times \sqrt{3}) = 2\sqrt{15}.$$

Example 2. Simplify. $3x\sqrt{3} \times 4\sqrt{x}$

Solution: Multiply the numbers outside of the radicals and the radical parts. Then, simplify:

$$3x\sqrt{3} \times 4\sqrt{x} = (3x \times 4) \times \left(\sqrt{3} \times \sqrt{x}\right) = (12x)\left(\sqrt{3x}\right) = 12x\sqrt{3x}.$$

Example 3. Evaluate. $6a\sqrt{7b} \times 3\sqrt{2b}$

Solution: Multiply the numbers outside of the radicals and the radical parts. Then:

$$6a\sqrt{7b} \times 3\sqrt{2b} = 6a \times 3 \times \sqrt{7b} \times \sqrt{2b} = 18a\sqrt{14b^2}.$$

Simplify:

$$18a\sqrt{14b^2} = 18a \times \sqrt{14} \times \sqrt{b^2} = 18ab\sqrt{14}.$$

Example 4. Simplify. $9\sqrt{9x} \times 5\sqrt{4x}$

Solution: Multiply the numbers outside of the radicals and the radical parts. Then, simplify: $9\sqrt{9x} \times 5\sqrt{4x} = (9 \times 5) \times \left(\sqrt{9x} \times \sqrt{4x}\right) = (45)\left(\sqrt{36x^2}\right) = 45\sqrt{36x^2}$.

$\sqrt{36x^2} = 6x$, then:

$45\sqrt{36x^2} = 45 \times 6x = 270x$.

Rationalizing Radical Expressions

- Radical expressions cannot be in the denominator. (Number in the bottom)
- To get rid of the radical in the denominator, multiply both the numerator and denominator by the radical in the denominator.
- If there is a radical and another integer in the denominator, multiply both the numerator and denominator by the conjugate of the denominator.
- The conjugate of $(a + b)$ is $(a - b)$ and vice versa.

Examples:

Example 1. Simplify: $\frac{9}{\sqrt{3}}$.

Solution: Multiply both the numerator and denominator by $\sqrt{3}$. Then:

$$\frac{9}{\sqrt{3}} \times \frac{\sqrt{3}}{\sqrt{3}} = \frac{9\sqrt{3}}{\sqrt{9}} = \frac{9\sqrt{3}}{3}.$$

Now, simplify: $\frac{9\sqrt{3}}{3} = 3\sqrt{3}$.

Example 2. Simplify $\frac{5}{\sqrt{6}-4}$.

Solution: Multiply by the conjugate: $\frac{\sqrt{6}+4}{\sqrt{6}+4} \rightarrow \frac{5}{\sqrt{6}-4} \times \frac{\sqrt{6}+4}{\sqrt{6}+4}$.

$$(\sqrt{6} - 4)(\sqrt{6} + 4) = -10,$$

then: $\frac{5}{\sqrt{6}-4} \times \frac{\sqrt{6}+4}{\sqrt{6}+4} = \frac{5(\sqrt{6}+4)}{-10}$.

Use the fraction rule:

$$\frac{a}{-b} = -\frac{a}{b} \rightarrow \frac{5(\sqrt{6}+4)}{-10} = -\frac{5(\sqrt{6}+4)}{10} = -\frac{1}{2}(\sqrt{6} + 4).$$

Example 3. Simplify $\frac{2}{\sqrt{3}-1}$.

Solution: Multiply by the conjugate: $\frac{\sqrt{3}+1}{\sqrt{3}+1}$.

$$\frac{2}{\sqrt{3}-1} \times \frac{\sqrt{3}+1}{\sqrt{3}+1} = \frac{2(\sqrt{3}+1)}{2} = (\sqrt{3} + 1).$$

Radical Equations

- Isolate the radical on one side of the equation.
- Square both sides of the equation to remove the radical.
- Solve the equation for the variable.
- Plugin the answer (answers) into the original equation to avoid extraneous values.

Examples:

Example 1. Solve $\sqrt{x} - 5 = 15$.

Solution: Add 5 to both sides:

$$(\sqrt{x} - 5) + 5 = 15 + 5 \rightarrow \sqrt{x} = 20,$$

square both sides:

$$(\sqrt{x})^2 = 20^2 \rightarrow x = 400.$$

Plugin the value of 400 for x in the original equation and check the answer:

$$x = 400 \rightarrow \sqrt{x} - 5 = \sqrt{400} - 5 = 20 - 5 = 15,$$

so, the value of 400 for x is correct.

Example 2. What is the value of x in this equation? $2\sqrt{x+1} = 4$

Solution: Divide both sides by 2. Then:

$$2\sqrt{x+1} = 4 \rightarrow \frac{2\sqrt{x+1}}{2} = \frac{4}{2} \rightarrow \sqrt{x+1} = 2.$$

Square both sides: $\left(\sqrt{(x+1)}\right)^2 = 2^2$.

Then $x + 1 = 4 \rightarrow x = 3$.

Substitute x by 3 in the original equation and check the answer:

$$x = 3 \rightarrow 2\sqrt{x+1} = 2\sqrt{3+1} = 2\sqrt{4} = 2(2) = 4.$$

So, the value of 3 for x is correct.

Domain and Range of Radical Functions

- To find the domain of the function, find all possible values of the variable inside the radical.
- Remember that having a negative number under the square root symbol is not possible. (For cubic roots, we can have negative numbers.)
- To find the range, plugin the minimum and maximum values of the variable inside radical.

Examples:

Example 1. Find the domain and range of the radical function.

$$y = \sqrt{x-3}$$

Solution: For domain: find non-negative values for radicals: $x - 3 \geq 0$.

Domain of functions: $\sqrt{f(x)} \rightarrow f(x) \geq 0$, then solve $x - 3 \geq 0 \rightarrow x \geq 3$.

Domain of the function $y = \sqrt{x-3}$: $x \geq 3$.

For range: the range of a radical function of the form $c\sqrt{ax+b} + k$ is:

$$f(x) \geq k$$

For the function $y = \sqrt{x-3}$, the value of k is 0. Then: $f(x) \geq 0$.

Range of the function $y = \sqrt{x-3}$: $f(x) \geq 0$.

Example 2. Find the domain and range of the radical function.

$$y = 6\sqrt{4x+8} + 5$$

Solution: For domain: find non-negative values for radicals: $4x + 8 \geq 0$.
Domain of functions: $4x + 8 \geq 0 \rightarrow 4x \geq -8 \rightarrow x \geq -2$.
Domain of the function $y = 6\sqrt{4x+8} + 5$: $x \geq -2$.
For range: the range of a radical function of the form $c\sqrt{ax+b} + k$ is: $f(x) \geq k$.
For the function $y = 6\sqrt{4x+8} + 5$, the value of k is 5. Then: $f(x) \geq 5$.
Range of the function $y = 6\sqrt{4x+8} + 5$: $f(x) \geq 5$.

Simplify Radicals with Fractions

- To simplify radicals with fractions:
  - Rewrite the numerator and denominator of the fraction as the product of the prime factorizations.
  - Apply the multiplication and division properties of radical expressions.
  - Group the factors that form a perfect square, perfect cube and etc.
  - Simplify.

Examples:

Example 1. Simplify. $\sqrt{\frac{9}{25}}$

Solution: To simplify the radical fraction, rewrite the numerator and denominator as the product of the prime factorizations. So, $\sqrt{\frac{9}{25}} = \sqrt{\frac{3\times3}{5\times5}}$.
Since the index of the given radical is 2. You can take one term out of radical for every two same terms multiplied inside the radical sign (perfect square). Then: $\sqrt{\frac{3\times3}{5\times5}} = \frac{3}{5}$.

Example 2. Simplify the following radical fraction: $\sqrt{\frac{44}{16}}$.

Solution: Rewrite the numerator and denominator of the fraction as follows: $\sqrt{\frac{44}{16}} = \sqrt{\frac{2\times2\times11}{2\times2\times2\times2}}$. Consider the index of the given radical, take out one term of radical for every term that is repeated in an even number inside the radical sign. So, $\sqrt{\frac{2\times2\times11}{2\times2\times2\times2}} = \frac{2\sqrt{11}}{4}$. Simplify, $\frac{2\sqrt{11}}{4} = \frac{\sqrt{11}}{2}$.

Example 3. Write the expression in the simplest radical form. $\sqrt{\frac{242}{45}}$

Solution: Rewrite this radical fraction as the product of the prime factorizations: $\sqrt{\frac{242}{45}} = \sqrt{\frac{2\times11\times11}{3\times3\times5}}$. Now, take out the terms that are perfect squares, so, $\sqrt{\frac{2\times11\times11}{3\times3\times5}} = \frac{11}{3}\sqrt{\frac{2}{5}}$.

Chapter 9: Practices

Evaluate.

1) $\sqrt{49} =$ ________
2) $\sqrt{4} \times \sqrt{81} =$ ________
3) $\sqrt{16} \times \sqrt{4x^2} =$ ________
4) $\sqrt{289} =$ ________
5) $\sqrt{25b^4} =$ ________
6) $\sqrt{9} \times \sqrt{x^2} =$ ________

Simplify.

7) $\sqrt{6} + 6\sqrt{6} =$
8) $9\sqrt{8} - 6\sqrt{2} =$
9) $-\sqrt{7} - 5\sqrt{7} =$
10) $10\sqrt{2} + 3\sqrt{18} =$
11) $\sqrt{12} - 6\sqrt{3} =$
12) $-2\sqrt{x} + 6\sqrt{x} =$

Evaluate.

13) $\sqrt{4} \times 2\sqrt{9} =$
14) $\sqrt{5} \times 3\sqrt{20y} =$
15) $-6\sqrt{4} \times 3\sqrt{4} =$
16) $-9\sqrt{3b^2} \times (-\sqrt{6}) =$

Simplify.

17) $\frac{1+\sqrt{5}}{1-\sqrt{3}} =$
18) $\frac{2+\sqrt{6}}{\sqrt{2}-\sqrt{5}} =$
19) $\frac{\sqrt{7}}{\sqrt{6}-\sqrt{3}} =$
20) $\frac{\sqrt{8a}}{\sqrt{a^5}} =$

Solve for x in each equation.

21) $2\sqrt{2x-4} = 8$
22) $9 = \sqrt{4x-1}$
23) $\sqrt{x} + 6 = 11$
24) $\sqrt{5x} = \sqrt{x+3}$

Effortless Math Education

Identify the domain and range of each function.

25) $y = \sqrt{x+1}$

26) $y = \sqrt{x-2} + 6$

27) $y = \sqrt{x} - 1$

28) $y = \sqrt{x-4}$

Simplify.

29) $\sqrt{\frac{625}{36}}$

30) $\sqrt{\frac{1296}{25}}$

31) $\sqrt{\frac{147}{64}}$

32) $\sqrt{\frac{98}{18}}$

Chapter 9: Answers

1) 7
2) 18
3) $8x$
4) 17
5) $5b^2$
6) $3x$
7) $7\sqrt{6}$
8) $12\sqrt{2}$
9) $-6\sqrt{7}$
10) $19\sqrt{2}$
11) $-4\sqrt{3}$
12) $4\sqrt{x}$
13) 12
14) $30y$
15) -72
16) $27b\sqrt{2}$
17) $-\frac{(1+\sqrt{5})(1+\sqrt{3})}{2}$
18) $-\frac{2\sqrt{2}+2\sqrt{5}+2\sqrt{3}+\sqrt{30}}{3}$
19) $\frac{\sqrt{7}(\sqrt{6}+\sqrt{3})}{3}$
20) $\frac{2\sqrt{2}}{a^2}$
21) $x = 10$
22) $x = 20.5$
23) $x = 25$
24) $x = \frac{3}{4}$
25) $x \geq -1, y \geq 0$
26) $x \geq 2, y \geq 6$
27) $x \geq 0, y \geq -1$
28) $x \geq 4, y \geq 0$
29) $\frac{25}{6}$
30) $\frac{36}{5}$
31) $\frac{7\sqrt{3}}{8}$
32) $\frac{7}{3}$

CHAPTER

10 Rational Expressions

Math topics that you'll learn in this chapter:

- ☑ Simplifying Complex Fractions
- ☑ Graphing Rational Functions
- ☑ Adding and Subtracting Rational Expressions
- ☑ Multiplying Rational Expressions
- ☑ Dividing Rational Expressions
- ☑ Evaluate Integers Raised to Rational Exponents

Simplifying Complex Fractions

- Convert mixed numbers to improper fractions.
- Simplify all fractions.
- Write the fraction in the numerator of the main fraction line then write the division sign (÷) and the fraction of the denominator.
- Use the normal method for dividing fractions.
- Simplify as needed.

Examples:

Example 1. Simplify: $\frac{\frac{3}{5}}{\frac{2}{25}-\frac{5}{16}}$.

Solution: First, simplify the denominator: $\frac{2}{25}-\frac{5}{16}=-\frac{93}{400}$, Then:

$$\frac{\frac{3}{5}}{\frac{2}{25}-\frac{5}{16}}=\frac{\frac{3}{5}}{(-\frac{93}{400})}.$$

Now, write the complex fraction using the division sign (÷): $\frac{\frac{3}{5}}{(-\frac{93}{400})}=\frac{3}{5}\div\left(-\frac{93}{400}\right)$.

Use the dividing fractions rule: Keep, Change, Flip (Keep the first fraction, Change the division sign to multiplication, Flip the second fraction)

$$\frac{3}{5}\div\left(-\frac{93}{400}\right)=\frac{3}{5}\times\left(-\frac{400}{93}\right)=-\frac{240}{93}=-\frac{80}{31}=-2\frac{18}{31}.$$

Example 2. Simplify: $\frac{\frac{2}{7}\div\frac{1}{4}}{\frac{7}{8}+\frac{1}{4}}$.

Solution: First, simplify the numerator: $\frac{2}{7}\div\frac{1}{4}=\frac{8}{7}$, then, simplify the denominator: $\frac{7}{8}+\frac{1}{4}=\frac{9}{8}$,

Now, write the complex fraction using the division sign (÷):

$$\frac{\frac{2}{7}\div\frac{1}{4}}{\frac{7}{8}+\frac{1}{4}}=\frac{\frac{8}{7}}{\frac{9}{8}}=\frac{8}{7}\div\frac{9}{8}.$$

Use the dividing fractions rule:

(Keep, Change, Flip) $\frac{8}{7}\div\frac{9}{8}=\frac{8}{7}\times\frac{8}{9}=\frac{64}{63}=1\frac{1}{63}$.

Graphing Rational Functions

- A rational expression is a fraction in which the numerator and/or the denominator are polynomials. Examples: $\frac{1}{x}, \frac{x^2}{x-1}, \frac{x^2-x+2}{x^2+5x+1}, \frac{m^2+6m-5}{m-2m}$.
- To graph a rational function:
  - Find the vertical asymptotes of the function if there are any. (Vertical asymptotes are vertical lines that correspond to the zeroes of the denominator. The graph will have a vertical asymptote at $x = a$ if the denominator is zero at $x = a$ and the numerator isn't zero at $x = a$.)
  - Find the horizontal or slant asymptote. (If the numerator has a bigger degree than the denominator, there will be a slant asymptote. To find the slant asymptote, divide the numerator by the denominator using either long division or synthetic division.)
  - If the denominator has a bigger degree than the numerator, the horizontal asymptote is the x −axes or the line $y = 0$. If they have the same degree, the horizontal asymptote equals the leading coefficient (The coefficient of the largest exponent) of the numerator divided by the leading coefficient of the denominator.
  - Find intercepts and plug in some values of x and solve for y, then graph the function.

Example:

Graph rational function. $f(x) = \frac{x^2-x+2}{x-1}$

Solution: First, notice that the graph is in two pieces. Most rational functions have graphs in multiple pieces. Find y −intercept by substituting zero for x and solving for y, $(f(x))$:

$x = 0 \rightarrow y = \frac{x^2-x+2}{x-1} = \frac{0^2-0+2}{0-1} = -2$,

y −intercept: $(0, -2)$

Asymptotes of $\frac{x^2-x+2}{x-1}$: Vertical: $x = 1$, Slant asymptote: $y = x$.

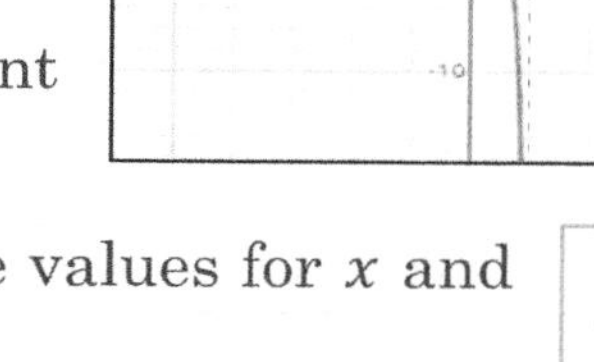

After finding the asymptotes, you can plug in some values for x and solve for y. Here is the sketch for this function.

Adding and Subtracting Rational Expressions

- For adding and subtracting rational expressions:
  - Find the least common denominator (LCD).
  - Write each expression using the LCD.
  - Add or subtract the numerators.
  - Simplify as needed.

Examples:

Example 1. Solve. $\frac{4}{2x+3}+\frac{x-2}{2x+3}=$

Solution: The denominators are equal. Then, use the fractions addition rule:

$\frac{a}{c}\pm\frac{b}{c}=\frac{a\pm b}{c}$.

Therefore:

$\frac{4}{2x+3}+\frac{x-2}{2x+3}=\frac{4+(x-2)}{2x+3}=\frac{x+2}{2x+3}$.

Example 2. Solve. $\frac{x+4}{x-5}+\frac{x-4}{x+6}=$

Solution: Find the least common denominator of $(x-5)$ and $(x+6)$:

$(x-5)(x+6)$.

Then:

$\frac{x+4}{x-5}+\frac{x-4}{x+6}=\frac{(x+4)(x+6)}{(x-5)(x+6)}+\frac{(x-4)(x-5)}{(x+6)(x-5)}=\frac{(x+4)(x+6)+(x-4)(x-5)}{(x+6)(x-5)}$.

Expand:

$(x+4)(x+6)+(x-4)(x-5)=2x^2+x+44$.

Then:

$\frac{(x+4)(x+6)+(x-4)(x-5)}{(x+6)(x-5)}=\frac{2x^2+x+44}{(x+6)(x-5)}=\frac{2x^2+x+44}{x^2+x-30}$.

Multiplying Rational Expressions

- Multiplying rational expressions is the same as multiplying fractions. First, multiply numerators and then multiply denominators. Then, simplify as needed.

Examples:

Example 1. Solve: $\frac{x+6}{x-1} \times \frac{x-1}{5} =$

Solution: Multiply numerators and denominators:

$$\frac{a}{b} \times \frac{c}{d} = \frac{a \times c}{b \times d}.$$

Therefore:

$$\frac{x+6}{x-1} \times \frac{x-1}{5} = \frac{(x+6)(x-1)}{5(x-1)}.$$

Cancel the common factor: $(x-1)$.

Then:

$$\frac{(x+6)(x-1)}{5(x-1)} = \frac{(x+6)}{5}.$$

Example 2. Solve: $\frac{x-2}{x+3} \times \frac{2x+6}{x-2} =$

Solution: Multiply numerators and denominators:

$$\frac{x-2}{x+3} \times \frac{2x+6}{x-2} = \frac{(x-2)(2x+6)}{(x+3)(x-2)}.$$

Cancel the common factor:

$$\frac{(x-2)(2x+6)}{(x+3)(x-2)} = \frac{(2x+6)}{(x+3)}.$$

Factor $2x + 6 = 2(x + 3)$.

Then: $\frac{2(x+3)}{(x+3)} = 2$.

Dividing Rational Expressions

- To divide rational expression, use the same method we use for dividing fractions. (Keep, Change, Flip)
- Keep the first rational expression, change the division sign to multiplication, and flip the numerator and denominator of the second rational expression. Then, multiply numerators and multiply denominators. Simplify as needed.

Examples:

Example 1. Solve. $\frac{x+2}{3x} \div \frac{x^2+5x+6}{3x^2+3x} =$

Solution: Use fractions division rule: $\frac{a}{b} \div \frac{c}{d} = \frac{a}{b} \times \frac{d}{c} = \frac{a \times d}{b \times c}$.

Therefore:

$$\frac{x+2}{3x} \div \frac{x^2+5x+6}{3x^2+3x} = \frac{x+2}{3x} \times \frac{3x^2+3x}{x^2+5x+6} = \frac{(x+2)(3x^2+3x)}{(3x)(x^2+5x+6)}.$$

Now, factorize the expressions $3x^2 + 3x$ and $(x^2 + 5x + 6)$. Then:

$3x^2 + 3x = 3x(+1)$ and $x^2 + 5x + 6 = (x + 2)(x + 3)$.

Simplify:

$$\frac{(x+2)(3x^2+3x)}{(3x)(x^2+5x+6)} = \frac{(x+2)(3x)(x+1)}{(3x)(x+2)(x+3)},$$

cancel common factors. Then:

$$\frac{(x+2)(3x)(x+1)}{(3x)(x+2)(x+3)} = \frac{x+1}{x+3}.$$

Example 2. Solve. $\frac{5x}{x+3} \div \frac{x}{2x+6} =$

Solution: Use fractions division rule: $\frac{a}{b} \div \frac{c}{d} = \frac{a}{b} \times \frac{d}{c} = \frac{a \times d}{b \times c}$.

Then:

$$\frac{5x}{x+3} \div \frac{x}{2x+6} = \frac{5x}{x+3} \times \frac{2x+6}{x} = \frac{5x(2x+6)}{x(x+3)} = \frac{5x \times 2(x+3)}{x(x+3)}.$$

Cancel common factor:

$$\frac{5x \times 2(x+3)}{x(x+3)} = \frac{10x(x+3)}{x(x+3)} = 10.$$

Evaluate Integers Raised to Rational Exponents

- In order to evaluate the rational exponents of integers, perform the following steps:

  Step 1: Write the base in exponential notations.

  Step 2: Multiply the exponents obtained and simplify.

Examples:

Example 1. Evaluate $81^{\frac{3}{2}}$.

Solution: First, rewrite the base as exponential form $81 = 3^4$. Now, we have:

$$81^{\frac{3}{2}} = (3^4)^{\frac{3}{2}}.$$

So, multiply two exponents, $(3^4)^{\frac{3}{2}} = 3^{4\times\frac{3}{2}} = 3^6 = 729$.

Example 2. Simplify $(-32)^{-\frac{1}{5}}$.

Solution: Write the base in exponential notation, $32 = 2^5$. Then:

$$(-32)^{-\frac{1}{5}} = (-2^5)^{-\frac{1}{5}}.$$

Use this formula $a^{-b} = \frac{1}{a^b}$, so: $(-2^5)^{-\frac{1}{5}} = \frac{1}{(-2^5)^{\frac{1}{5}}} = \frac{1}{(-2)^{5\times\frac{1}{5}}} = -\frac{1}{2}$.

Example 3. Calculate the value of $\sqrt[3]{121^{\frac{3}{2}}}$.

Solution: Use this formula $\sqrt[n]{a} = a^{\frac{1}{n}}$. Then:

$$\sqrt[3]{121^{\frac{3}{2}}} = \left(121^{\frac{3}{2}}\right)^{\frac{1}{3}}.$$

Now, by using $(a^n)^m = a^{nm}$. So, we have: $\left(121^{\frac{3}{2}}\right)^{\frac{1}{3}} = 121^{\frac{3}{2}\times\frac{1}{3}} = 121^{\frac{1}{2}}$.

Write the base in exponential notation, $121^{\frac{1}{2}} = (11^2)^{\frac{1}{2}}$. Multiply the exponents and simplify: $(11^2)^{\frac{1}{2}} = 11^{2\times\frac{1}{2}} = 11$.

Chapter 10: Practices

Simplify each expression.

1) $\frac{\frac{8}{3}}{\frac{2}{5}} =$

2) $\frac{\frac{x}{3}+\frac{x}{8}}{\frac{1}{4}} =$

3) $\frac{\frac{x+3}{3}}{\frac{x-2}{2}} =$

4) $\frac{1+\frac{x}{4}}{x} =$

Graph rational expressions.

5) $f(x) = \frac{x^2-2x}{x-3}$

6) $f(x) = \frac{6x+1}{x^2-4x}$

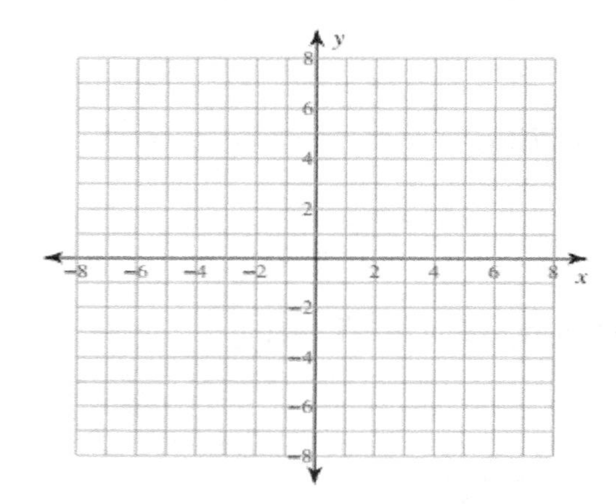

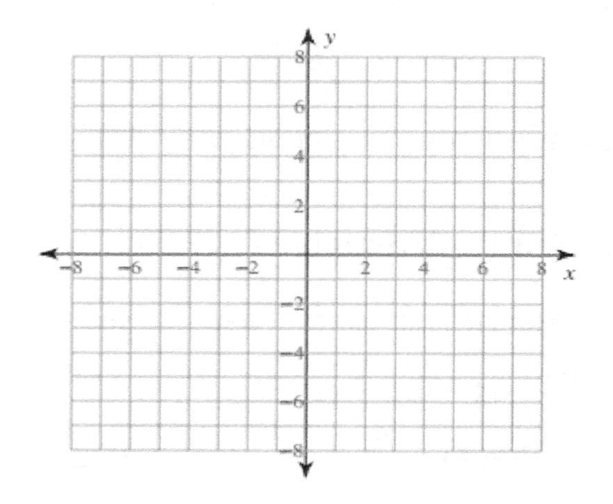

Simplify each expression.

7) $\frac{x+6}{x+1} - \frac{x+9}{x+1} =$

8) $\frac{2x+1}{x+3} + \frac{2}{x+4} =$

9) $\frac{14}{x+4} + \frac{6}{x^2-16} =$

10) $\frac{x+2}{x+8} - \frac{2x}{x-8} =$

Simplify each expression.

11) $\frac{20x^3}{3} \times \frac{15}{4x} =$

12) $\frac{x+6}{4} \times \frac{16}{x+6} =$

13) $\frac{x+10}{4x} \times \frac{3x}{7x+70} =$

14) $\frac{x+8}{x+6} \times \frac{x-6}{4x+32} =$

Simplify each expression.

15) $\frac{10x}{x+2} \div \frac{x}{60x+120} =$

16) $\frac{5}{4} \div \frac{45}{8x} =$

17) $\frac{x-6}{x+3} \div \frac{4}{x+3} =$

18) $\frac{7x^3}{x^2-64} \div \frac{x^3}{x^2+x-56} =$

Simplify.

19) $(121x^6)^{\frac{1}{2}}$

20) $(64x^{12})^{\frac{1}{6}}$

21) $(-32)^{-\frac{1}{5}}$

22) $(-27)^{\frac{1}{3}}$

Effortless Math Education

Chapter 10: Answers

1) $\frac{20}{3}$

2) $\frac{11x}{6}$

3) $\frac{2x+6}{3x-6}$

4) $\frac{4+x}{4x}$

5) $f(x)=\frac{x^2-2x}{x-3}$

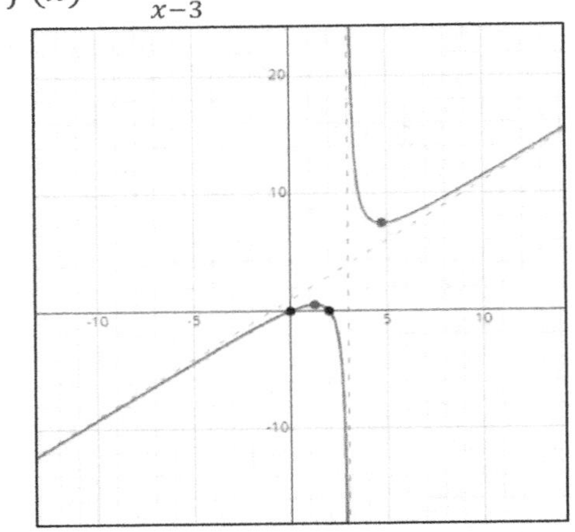

6) $f(x)=\frac{6x+1}{x^2-4x}$

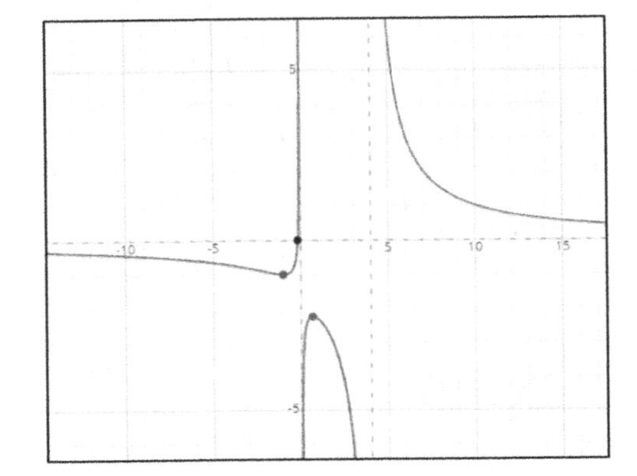

7) $-\frac{3}{x+1}$

8) $\frac{2x^2+11x+10}{(x+3)(x+4)}$

9) $\frac{14x-50}{(x+4)(x-4)}$

10) $\frac{-x^2-22x-16}{(x+8)(x-8)}$

11) $25x^2$

12) 4

13) $\frac{3}{28}$

14) $\frac{x-6}{4(x+6)}$

15) 600

16) $\frac{2x}{9}$

17) $\frac{x-6}{4}$

18) $\frac{7(x-7)}{x-8}$

19) $11x^3$

20) $2x^2$

21) $-\frac{1}{2}$

22) -3

CHAPTER

11 Statistics and Probabilities

Math topics that you'll learn in this chapter:

- ☑ Mean, Median, Mode, and Range of the Given Data
- ☑ Pie Graph
- ☑ Scatter Plots
- ☑ Probability Problems
- ☑ Permutations and Combinations
- ☑ Calculate and Interpret Correlation Coefficients
- ☑ Equation of a Regression Line and Interpret Regression Lines
- ☑ Correlation and Causation

Mean, Median, Mode, and Range of the Given Data

- **Mean:** $\frac{sum\ of\ the\ data}{total\ number\ of\ data\ entires}$
- **Mode:** the value in the list that appears most often
- **Median:** is the middle number of a group of numbers arranged in order by size.
- **Range:** the difference between the largest value and smallest value in the list

Examples:

Example 1. What is the mode of these numbers? 5, 6, 8, 6, 8, 5, 3, 5

Solution: Mode: the value in the list that appears most often.

Therefore, the mode is number 5. There are three number 5 in the data.

Example 2. What is the median of these numbers? 6, 11, 15, 10, 17, 20, 7

Solution: Write the numbers in order: 6, 7, 10, 11, 15, 17, 20

The median is the number in the middle. Therefore, the median is 11.

Example 3. What is the mean of these numbers? 7, 2, 3, 2, 4, 8, 7, 5

Solution: Use this formula:

$$\text{Mean: } \frac{sum\ of\ the\ data}{total\ number\ of\ data\ entires}.$$

Therefore:

$$\text{Mean: } = \frac{7+2+3+2+4+8+7+5}{8} = \frac{38}{8} = 4.75$$

Example 4. What is the range in this list? 3, 7, 12, 6, 15, 20, 8

Solution: The range is the difference between the largest value and the smallest value in the list. The largest value is 20 and the smallest value is 3. Then:

$20 - 3 = 17$.

Pie Graph

- A Pie Graph (Pie Chart) is a circle chart divided into sectors; each sector represents the relative size of each value.
- Pie charts represent a snapshot of how a group is broken down into smaller pieces.

Examples:

A library has 750 books that include Mathematics, Physics, Chemistry, English and History. Use the following graph to answer the questions.

Example 1. What is the number of Mathematics books?

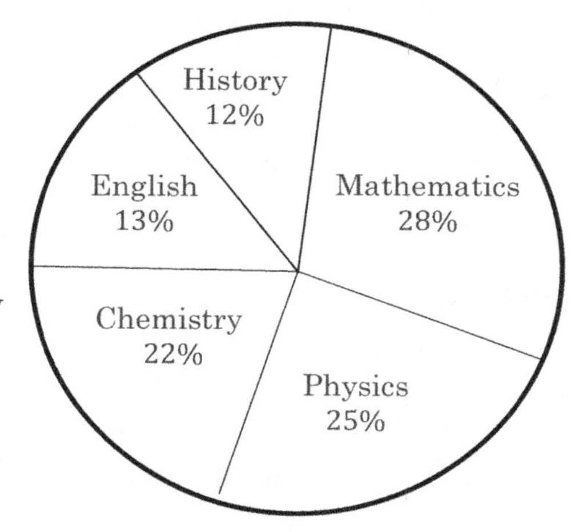

Solution: Number of total books $= 750$.
Percent of Mathematics books $= 28\%$.
Then, the number of Mathematics books:

$$28\% \times 750 = 0.28 \times 750 = 210.$$

Example 2. What is the number of History books?

Solution: Number of total books $= 750$.
Percent of History books $= 12\%$.
Then:

$$0.12 \times 750 = 90.$$

Example 3. What is the number of Chemistry books in the library?

Solution: Number of total books $= 750$.
Percent of Chemistry books $= 22\%$.
Then:

$$0.22 \times 750 = 165.$$

Scatter Plots

- A scatter plot is a diagram with points to represent the relationship between two variables.
- On a scatter plot, you can use a trend line to make predictions.
- A scatter plot shows a positive trend if y tends to increase as x increases.
- A scatter plot shows a negative trend if y tends to decrease as x increases.
- An outlier is an extreme point in a data set that is separated from all other points.

Example:

The following table shows the number of people in a family and the amount of money they spend on movie tickets.

Number of people	1	2	3	4	5	6	7
Money ($)	13	14	17	15	28	18	16

a) Make a scatter plot to represent the data.

b) Does this scatter plot show a positive trend, a negative trend, or no trend?

c) Find the outlier on the scatter plot.

Solution:

a) Write the ordered pairs. Number of people goes on the x −axis, so put the number of people first. The amount of money goes on the y −axis, so put the amount of money second. (1,13), (2,14), (3,17), (4,15), (5,28), (6,18), (7,16).

Now, graph the ordered pairs.

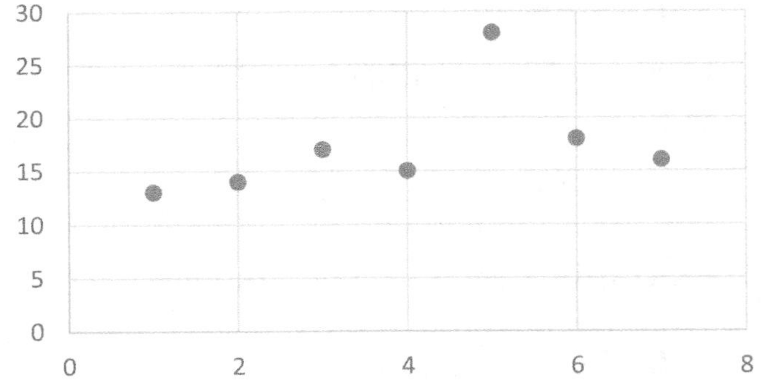

b) y tends to increase as x increases. So, this scatter plot shows a positive trend.

c) (5,28) is the outlier because this point is separated from all other points in the data set.

Probability Problems

- Probability is the likelihood of something happening in the future. It is expressed as a number between zero (Can never happen) to 1 (Will always happen).
- Probability can be expressed as a fraction, a decimal, or a percent.
- Probability formula: $Probability = \frac{number\ of\ desired\ outcomes}{number\ of\ total\ outcomes}$.

Examples:

Example 1. Anita's trick–or–treat bag contains 10 pieces of chocolate, 16 suckers, 16 pieces of gum, 22 pieces of licorice. If she randomly pulls a piece of candy from her bag, what is the probability of her pulling out a piece of sucker?

Solution: Use this formula:

$$Probability = \frac{number\ of\ desired\ outcomes}{number\ of\ total\ outcomes}.$$

Probability of pulling out a piece of sucker $= \frac{16}{10+16+16+22} = \frac{16}{64} = \frac{1}{4}$.

Example 2. A bag contains 20 balls: four green, five black, eight blue, one brown, one red and one white. If 19 balls are removed from the bag at random, what is the probability that a brown ball has been removed?

Solution: If 19 balls are removed from the bag at random, there will be one ball in the bag. The probability of choosing a brown ball is 1 out of 20. Therefore, the probability of not choosing a brown ball is 19 out of 20 and the probability of having not a brown ball after removing 19 balls is the same.

The answer is: $\frac{19}{20}$.

Permutations and Combinations

- **Factorials** are products, indicated by an exclamation mark. For example, $4! = 4 \times 3 \times 2 \times 1$. (Remember that 0! is defined to be equal to 1.)
- **Permutations:** The number of ways to choose a sample of k elements from a set of n distinct objects where order does matter, and replacements are not allowed. For a permutation problem, use this formula:

$$nP_k = \frac{n!}{(n-k)!}$$

- **Combination:** The number of ways to choose a sample of r elements from a set of n distinct objects where order does not matter, and replacements are not allowed. For a combination problem, use this formula:

$$nC_r = \frac{n!}{r!(n-r)!}$$

Examples:

Example 1. How many ways can the first and second place be awarded to 7 people?

Solution: Since the order matters, (The first and second place are different!) we need to use a permutation formula where n is 7 and k is 2.

Then:

$$\frac{n!}{(n-k)!} = \frac{7!}{(7-2)!} = \frac{7!}{5!} = \frac{7 \times 6 \times 5!}{5!},$$

remove 5! from both sides of the fraction.

Then:

$$\frac{7 \times 6 \times 5!}{5!} = 7 \times 6 = 42.$$

Example 2. How many ways can we pick a team of 3 people from a group of 8?

Solution: Since the order doesn't matter, we need to use a combination formula where n is 8 and r is 3.

Then:

$$\frac{n!}{r!(n-r)!} = \frac{8!}{3!(8-3)!} = \frac{8!}{3!(5)!} = \frac{8 \times 7 \times 6 \times 5!}{3!(5)!} = \frac{8 \times 7 \times 6}{3 \times 2 \times 1} = \frac{336}{6} = 56.$$

Calculate and Interpret Correlation Coefficients

- The correlation coefficient which is represented with sign r determines how close the points of a data set are to being linear. In other words, it evaluates the linear correlation's strength in the data set. The correlation coefficient is used for a set of n data points, (x_i, y_i) where $1 \leq i \leq n$. Whenever the correlation coefficient is closer to 1 or -1, the linear correlation of the data can be stronger. The correlation coefficient is 1 if the points of the data set are on a line with a positive slope. The correlation coefficient is -1 if the data set's points are on a line with a negative slope.
- The formula for the correlation coefficient is as follows $r = \frac{1}{n-1} \cdot \sum_{i=1}^{n} \frac{(x_i - \bar{x})(y_i - \bar{y})}{s_x s_y}$. In this formula, $\bar{x}$ is the $x-$values' mean, $\bar{y}$ is the $y-$values' mean, s_x is the $x-$values' sample standard deviation, s_y is the $y-$values' sample standard deviation, and n is the data points' number.
- If the data points have a positive trend or an increasing trend, the correlation coefficient is positive. If the data points have a negative trend or decreasing trend, the correlation coefficient is negative.

Example:

Find the correlation coefficient of the following data and then interpret the answer. Round your answer to the nearest thousandth.

Arthur plans to improve his studies. In order to achieve this goal, he has written down the number of pages read daily and the hours of daily reading in the table below. He records the hours of daily reading, x, and the number of pages reads daily, y.

Hours of daily reading	The number of pages read daily
3	60
4	75
6	82
7	90
8	105

Solution: Each row in the above table shows a data point (x_i, y_i). x_i is the number of hours Arthur reads per day and y_i is the number of pages Arthur reads per day. You can simply use your calculator to find the above data set's correlation coefficient: $r = 0.968971 \approx 0.969$. The correlation coefficient is positive. So, the data points have a positive trend or increasing trend.
The correlation coefficient is so close to 1, so the data set has a strong linear correlation.

Equation of a Regression Line and Interpret Regression Lines

- The association between scattered data points in any set can be shown with a regression line. In fact, a regression line is a single line that best fits the data and has the least distance from the points. The regression line is used to forecast values based on a data set. The differences between the given values and the values forecasted by the regression line are called residuals. By looking at residuals you can see whether a line fits the data well or not.
- The regression line's formula is the same as a line formula in algebra ($y = mx + b$). In this formula, m is the regression line slope and b is the y-intercept.
- A line where the sum of the residuals' squares is minimized is called the least squares regression line and its formula is as follows: $y = ax + b$. The least squares regression line is usually used to make forecasts for a data set.
- Remember you should calculate the following values before calculating a regression line (you can also use a scientific calculator to find the answer):
    - The x −values' mean
    - The y −values' mean
    - The x −values' standard deviation
    - The y −values' standard deviation
    - The correlation between x and y

Example:

Find the equation for the least squares' regression line of the following data set. Round your answers to the nearest hundredth. Then interpret the regression line.

x_i	y_i
9	24
15	20
23	17
32	13
45	9

Solution: Use a calculator to find the least squares regression line's equation for the above data set: $y = -0.4085x + 26.73$. Round your answers to the nearest hundredth: $y = -0.41x + 26.73$. The slope is negative so, there is a negative linear relationship. It means when one variable increases the other variable decreases.

Correlation and Causation

- Correlation explains a connection between variables. When a variable changes the other variable also changes. In other words, a correlation is a statistical indicator of the association between variables. These variables have covariation and change together. But remember this covariation isn't certain because of a direct or indirect causal connection.
- Causation shows you when one variable changes, it causes changes in the other variable. In fact, you can see a cause-and-effect relationship between variables. The 2 variables are associated together and there is also a causal connection between them.
- Keep in mind a correlation doesn't signify causation, but causation always signifies correlation. There are 2 situations when a correlation isn't causation:

  - The third variable problem happens when a confounded variable affects two other variables and makes them seem causally linked when they are not.
  - The directionality problem happens when 2 variables have a correlation connection and might really have a causal link, but it isn't possible to determine which variable is the reason for changing in the other.

Examples:

Example 1. Determine whether the following relationship reflects both correlation and causation or not.

For book readers, having more free time is associated with reading more books.

Solution: This relationship reflects both correlation and causation because when a variable changes the other variable also changes. On the other hand, when one variable changes, it causes changes in the other variable. Naturally, for book readers, having more free time will lead to reading more books.

Example 2. Determine whether the following relationship reflects both correlation and causation or not.

Comedy shows on TV are associated with reading more about comedians.

Solution: This relationship doesn't reflect both correlation and causation because reading about comedians doesn't cause people to watch comedy shows on TV. Another factor should be used to explain the correlation, like being interested in comedy and comedians.

Chapter 11: Practices

Find the values of the given data.

1) 6, 11, 5, 3, 6
Mode: _____ Range: _____
Mean: _____ Median: _____

2) 4, 9, 1, 9, 6, 7
Mode: _____ Range: _____
Mean: _____ Median: _____

3) 10, 3, 6, 10, 4, 15
Mode: _____ Range: _____
Mean: _____ Median: _____

4) 12, 4, 8, 9, 3, 12, 15
Mode: _____ Range: _____
Mean: _____ Median: _____

The circle graph below shows all of Bob's expenses for last month. Bob spent $790 on his Rent last month.

5) How much did Bob's total expenses last month? ________

6) How much did Bob spend for foods last month? ________

7) How much did Bob spend for his bills last month? ________

8) How much did Bob spend on his car last month? ________

Bob's last month expenses

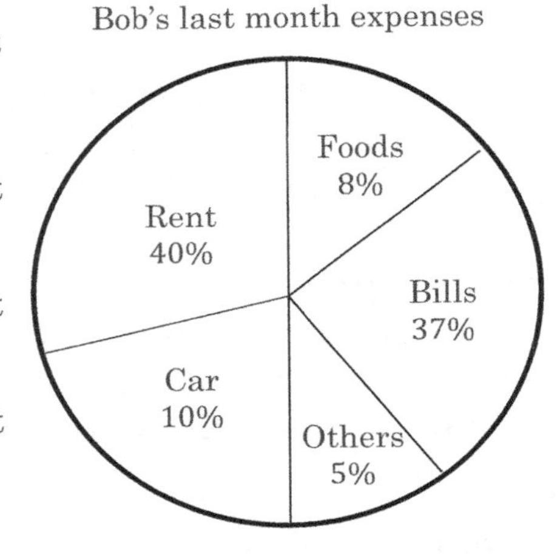

Make a scatter plot of the data.

9) Does this scatter plot show a positive trend, a negative trend, or no trend?

hours	amount
1	$12
2	$8
3	$14
4	$5
5	$7
6	$10
7	$4
8	$12

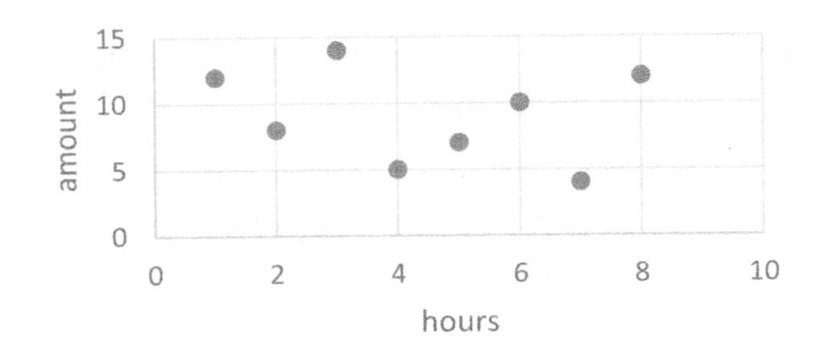

Solve.

10) Bag A contains 8 red marbles and 6 green marbles. Bag B contains 5 black marbles and 7 orange marbles. What is the probability of selecting a green marble at random from bag A? What is the probability of selecting a black marble at random from Bag B?

________________ ________________

Solve.

11) Susan is baking cookies. She uses sugar, flour, butter, and eggs. How many different orders of ingredients can she try? ________________
12) Jason is planning for his vacation. He wants to go to a museum, go to the beach, and play volleyball. How many different ways of ordering are there for him? ________________
13) In how many ways can a team of 6 basketball players choose a captain and co-captain? ________________
14) How many ways can the first and second place be awarded to 11 people? _____
15) A professor is going to arrange her 5 students in a straight line. In how many ways can she do this? ________________
16) In how many ways can a teacher choose 12 out of 15 students?

____

Find the correlation coefficient of the following data.

17)

x	12	14	18	21	28
y	2	4	6	8	12

18)

x	50	51	52	53	54
y	4.1	4.2	4.3	4.4	4.5

Effortless Math Education

Determine the linear regression equation from the given set of data.

19)

x	2	3	5	8
y	3	6	4	13

20)

x	2	4	6	8
y	4	7	10	12

Determine whether the following relationships reflect both correlation and causation or not.

21) The number of cold and snowy days and the amount of coffee at the ski resort.

22) The number of miles traveled, and the gas used.

Chapter 11: Answers

1) Mode: 6, Range: 8,
   Mean: 6.2, Median: 6
2) Mode: 9, Range: 8,
   Mean: 6, Median: 6.5
3) Mode: 10, Range: 12,
   Mean: 8, Median: 8
4) Mode: 12, Range: 12,
   Mean: 9, Median: 9
5) $1,975
6) $158
7) $730.75
8) $197.50
9) No trend
10) $\frac{3}{7}, \frac{5}{12}$
11) 24
12) 6
13) 30
14) 110
15) 120
16) 455
17) 0.997
18) 1
19) $y = 1.47x - 0.14$
20) $y = 1.35x + 1.5$
21) Correlation no causation
22) Correlation and causation

CHAPTER

12 Direct and Inverse Variation

Math topics that you'll learn in this chapter:

- ☑ Find the Constant of Variation
- ☑ Model Inverse Variation
- ☑ Write and Solve Direct Variation Equations

Find the Constant of Variation

- Variation signifies how one variable can be modified in mathematics related to another variable. Variation is generally represented as a ratio. When the variation is constant it means that the ratio of variation remains with no change. So, the constant variation in the relationship between the variables doesn't change.
- Constant variation is usually seen in one of these two types of formats: direct variation or inverse variation. In both of these formats, k is considered the constant of variation. If you have an equation in one of these two formats, you can easily find the constant of variation:

$$Inverse\ Variation \rightarrow k = xy$$

$$Direct\ Variation \rightarrow k = \frac{y}{x}$$

- Since in variation relations, k is constant and the same for every point, you can determine k when you have any point by division of the y −coordinate by the x −coordinate.
- Note that all equations don't have a constant of variation and they can't be written this way.

Examples:

Example 1. If y varies directly as x, and $x = 24$ when $y = 21$, what is the direct variation equation?

Solution: First, put x and y values in the $Direct\ Variation \rightarrow k = \frac{y}{x}$ to find the constant of variation: $Direct\ Variation \rightarrow k = \frac{y}{x} \rightarrow k = \frac{21}{24} \rightarrow k = \frac{7}{8}$. Now using the constant of variation, write the direct variation equation: $y = \frac{7}{8}x$.

Example 2. If y varies inversely as x, and $x = 4$ when $y = 6$, what is the inverse variation equation?

Solution: First, put x and y values in the $Inverse\ Variation \rightarrow k = xy$ to find the constant of variation: $Inverse\ Variation \rightarrow k = xy \rightarrow k = 4 \times 6 \rightarrow k = 24$. Now using the constant of variation, write the inverse variation equation: $y = \frac{24}{x}$.

Example 3. If y varies directly as x, and $y = 20$ when $x = 14$, then what is y when $x = 7$?

Solution: First, use x and y values to find the constant of variation: $Direct\ Variation \rightarrow k = \frac{y}{x} \rightarrow k = \frac{20}{14} \rightarrow k = \frac{10}{7}$. Now using the constant of variation, write the direct variation equation: $y = \frac{10}{7}x$. Find y when $x = 7 \rightarrow y = \frac{10}{7}x \rightarrow y = \frac{10}{7} \times 7 \rightarrow y = \frac{70}{7} = 10$.

Model Inverse Variation

A sample of an Inverse Variation graph

- The inverse variation is a kind of connection between variables that are shown in the form of $y = \frac{k}{x}$. In this relationship, x and y are 2 variables and k is a constant value. It shows that when the value of one number increases, the value of the other number changes inversely and decreases. The value of k remains unchanged and it can't be zero ($k \neq 0$).
- You can make the table of inverse variation by putting the one quantity's values in the equation and table, then find the other quantity.
- If you have an inverse variation and want to solve for an unknown, you can follow these steps:
  - Determine x, and y (the inputs and the outputs).
  - Find the variation's constant. Sometimes you should multiply y by a particular x power to find the variation's constant.
  - Write an equation for the relationship by using the constant of variation.
  - Find the unknown value by substitution known values into the equation.

Example:

According to the following table, determine if y values change inversely with x. If yes, write an equation for the inverse variation and show it in a graph.

x	y
3	16
4	12
6	8
8	6

Solution: In a table with inverse changes the product of all x and y pairs of its data is the same value. You can see that the product of any pair of x and y is equal to 48, so $k = 48$. Write an equation for the inverse variation:

$y = \frac{k}{x} \rightarrow y = \frac{48}{x}$

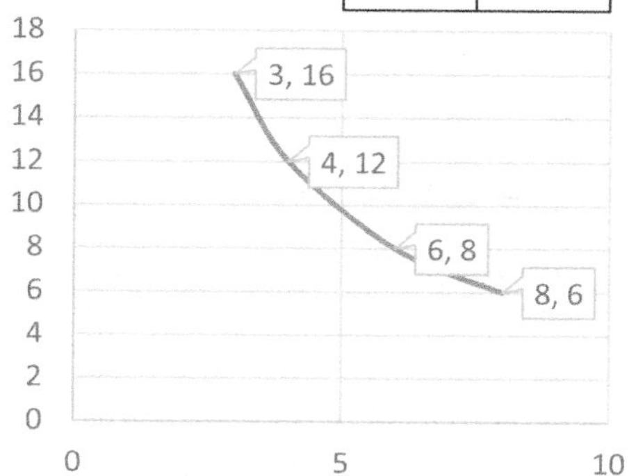

You can make a graph of the equation $y = \frac{48}{x}$ with points from the table as follows:

bit.ly/3IuhUYW

Find more at

Write and Solve Direct Variation Equations

- Direct variation expresses a math relationship between 2 variables. It shows y changes directly with x and whenever x increases, y increases, and conversely, whenever x decreases, y decreases. Keep in your mind that the ratio in a direct variation always remains the same.
- A two-variables linear equation's slope-intercept form is as follows: $y = mx + b$. Now if you consider the $y-$intercept occurs at the point of (0,0) and substitute the slope (m) with the constant of variation (k), you get a direct variation equation: $y = kx$. k is the constant of variation and it's a value that always remains with no changes.
- In the direct variation, the proportion between the variables stays the same, therefore the direct variation graph always makes a straight line.

Examples:

Example 1. If y varies directly with x and $y = 21$ when $x = 7$, find y when $x = 8$.

Solution: First, you should find the constant of variation. Put $x = 7$ and $y = 21$ into the direct variation equation to find the value of k: $y = kx \rightarrow 21 = k \times 7 \rightarrow k = \frac{21}{7} = 3 \rightarrow k = 3$. Now, use $k = 3$ to find y when $x = 8$: $y = kx \rightarrow y = 3 \times 8 = 24 \rightarrow y = 24$.

Example 2. If y varies directly with x and $y = 16$ when $x = 4$, find x when $y = 32$.

Solution: First, you should find the constant of variation. Put $x = 4$ and $y = 16$ into the direct variation equation to find the value of k: $y = kx \rightarrow 16 = k \times 4 \rightarrow k = \frac{16}{4} = 4 \rightarrow k = 4$. Now, use $k = 4$ to find x when $y = 32$: $y = kx \rightarrow 32 = 4 \times x = x = \frac{32}{4} = 8 \rightarrow x = 8$.

Example 3. If y varies directly with x and $y = 42$ when $x = 6$, find y when $x = 11$.

Solution: First, you should find the constant of variation. Put $x = 6$ and $y = 42$ into the direct variation equation to find the value of k: $y = kx \rightarrow 42 = k \times 6 \rightarrow k = \frac{42}{6} = 7 \rightarrow k = 7$. Now, use $k = 7$ to find y when $x = 11$: $y = kx \rightarrow y = 7 \times 11 = 77 \rightarrow y = 77$.

bit.ly/3EQww1p

Find more at

Chapter 12: Practices

Solve.

1) Let x and y be in direct variation $x = 6$ and $y = 22$. Then find the direct variation equation.
2) If $x = 15$ and $y = 30$ follow a direct variation, then find the constant of proportionality.
3) If y varies inversely as x and $x = 5$ when $y = 7$, what is the inverse variation equation?
4) Tell whether y varies inversely with x in the table below. If yes, write an equation for the inverse variation and show it in a graph.

x	y
2	12
4	6
6	4
8	3

5) If y varies directly with x and $y = 8$ when $x = 12$, find y when $x = -6$.
6) If y varies directly with x, find the missing value of x in $(-3, 27)$ and $(x, -27)$.
7) If y varies directly as x and $y = 12$ when $x = 2$, find y when $x = 8$.

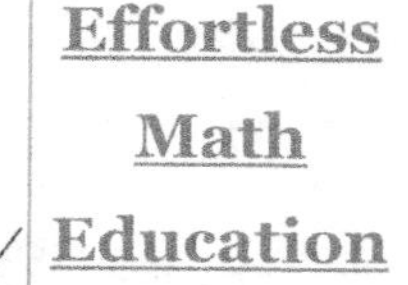

Chapter 12: Answers

1) $y = \frac{11}{3}x$

2) $k = 2$

3) $y = \frac{35}{x}$

4) $y = \frac{24}{x}$

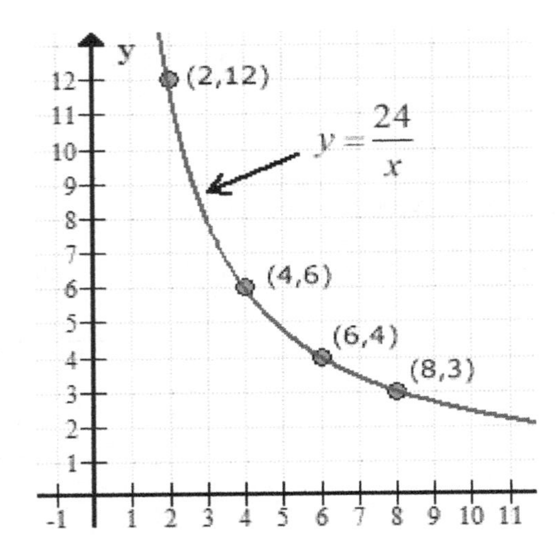

5) $y = -4$

6) $x = 3$

7) $y = 48$

CHAPTER

13 Number Sequences

Math topics that you'll learn in this chapter:

- ☑ Evaluate Recursive Formulas for Sequences
- ☑ Evaluate Variable Expressions for Number Sequences
- ☑ Write Variable Expressions for Arithmetic Sequences
- ☑ Write Variable Expressions for Geometric Sequences
- ☑ Write a Formula for a Recursive Sequence

Evaluate Recursive Formulas for Sequences

- Let $T_1, T_2, \cdots, T_n, \cdots$ is a sequence. T_n denotes the general term of the sequence.
- The sequence whose general term defines each term of a sequence using a previous term or terms is called a recursive sequence. The general formula is as follows:

$$T_{n+1} = f(T_n)$$

- To find the terms of a recursive sequence, according to the recursive formula, a number of previous terms of the sequence are required.

Examples:

Example 1. Find a_2 and a_3, where the first term is $a_1 = -1$, and the general term is $a_n = 2a_{n-1} - 1$.
Solution: Use the recursive sequence $a_n = 2a_{n-1} - 1$. To find a_2, plug $n = 2$ in $a_n = 2a_{n-1} - 1$. So, $a_2 = 2a_{2-1} - 1 \rightarrow a_2 = 2a_1 - 1$. Now, substitute $a_1 = -1$ into the obtained formula $a_2 = 2a_1 - 1$. Therefore, $a_2 = 2(-1) - 1 = -3$. In the same way, plug in $n = 3$. Then, $a_3 = 2a_{3-1} - 1 \rightarrow a_3 = 2a_2 - 1$. To find the value of a_3, put the obtained value for a_2 in this relation $a_3 = 2a_2 - 1$. Therefore, $a_2 = -3 \rightarrow a_3 = 2(-3) - 1 = -7$.
The terms are $a_2 = -3$ and $a_3 = -7$.
Example 2. Write the first five terms of the sequence $T_n = T_{n-1} + T_{n-2}$, where $n \geq 3$, $T_1 = 1$ and $T_2 = 3$.
Solution: Recall that the recursive formula defines each term of a sequence using a previous term or terms, then by looking at the recursive formula $T_n = T_{n-1} + T_{n-2}$, you notice that to generate the terms of the sequence, you need the previous two terms of the sequence. Here you are given the first two terms $T_1 = 1$ and $T_2 = 3$ together with the recursive formula $T_n = T_{n-1} + T_{n-2}$.
To find the third term which is T_3, plug in $n = 3$. So, $T_3 = T_{3-1} + T_{3-2} \rightarrow T_3 = T_2 + T_1$. Now, substitute the previous terms into the obtained expression. That is $T_3 = 3 + 1 = 4$. Similarly, by substituting $n = 4$ and $n = 5$ into the given recursive formula, you get:

$$T_4 = T_{4-1} + T_{4-2} \rightarrow T_4 = T_3 + T_2 \rightarrow T_4 = 4 + 3 = 7,$$
$$T_5 = T_{5-1} + T_{5-2} \rightarrow T_5 = T_4 + T_3 \rightarrow T_5 = 7 + 4 = 11.$$

The first five terms of the sequence are 1, 3, 4, 7, and 11.

Evaluate Variable Expressions for Number Sequences

- A sequence is an ordered list of numbers.
- Let $x_1, x_2, \cdots, x_i, \cdots$ represent a sequence. Each number in the sequence is called a term. The expression x_i is referred to the general term of the sequence. x_1 is the first term, x_2 is the second term, and so on. The three dots mean to continue forward in the pattern established. The subscript i on the term x_i is called index.
- In order to recognize the terms of the sequence, the general term of the sequence is used. In the sequences whose general term is only dependent on the variable (index value), by plugging the index in the general term, other values of the sequence are obtained.

Examples:

Example 1. Find the first three terms of the sequence with the general term $a_n = (-1)^{n-1}n$, where n represents the position of a term in the sequence and $n \geq 1$.

Solution: Since $n \geq 1$, the sequence starts at one. So, to find the first term, plug-in $n = 1$. Then: $a_1 = (-1)^{1-1} \times 1 = 1$. In a similar way, enter the natural numbers in order to find the other terms of the sequence. Therefore,

Plug in $n = 2$ as $a_2 = (-1)^{2-1} \times 2 = -2$.

Plug in $n = 3$ as $a_3 = (-1)^{3-1} \times 3 = 3$.

The first three terms of the sequence are 1, −2, and 3.

Example 2. Find the first five terms of the sequence with the general term $x_k = \left(-\frac{1}{2}\right)^k$, where k represents the position of a term in the sequence and starts with $k = 1$.

Solution: To find the first term, plug-in $k = 1$. Then: $x_1 = \left(-\frac{1}{2}\right)^1 = -\frac{1}{2}$. In the same way, to find the 2nd term, plug-in $k = 2$. So, $x_2 = \left(-\frac{1}{2}\right)^2 = \frac{1}{4}$. To find the 3rd term, plug-in $k = 3 \rightarrow x_3 = \left(-\frac{1}{2}\right)^3 = -\frac{1}{8}$. Finally, $k = 4 \rightarrow x_4 = \left(-\frac{1}{2}\right)^4 = \frac{1}{16}$, and $k = 5 \rightarrow x_5 = \left(-\frac{1}{2}\right)^5 = -\frac{1}{32}$.

The first three terms of the sequence are $-\frac{1}{2}, \frac{1}{4}, -\frac{1}{8}, \frac{1}{16}$, and $-\frac{1}{32}$.

bit.ly/3ZV4MSr
Find more at

Write Variable Expressions for Arithmetic Sequences

- The given recursive sequence x_1, x_2, x_3, $\cdots$ such that the difference of the terms of the sequence is equal to the same value, is called an arithmetic sequence.
- To write the variable expression for the arithmetic sequences, follow these steps:
  - Specify the initial values as $x_1 = a$.
  - Determine $d =$ the common difference between terms.
  - Add the first term with the product of the number of terms in the common difference.

Examples:

Example 1. Write the variable expression for the following sequence 3, 6, 9, 12, 15, $\cdots$.

Solution: According to the sequence, the common difference between the consecutive terms is $d = 3$. By dividing the terms by 3, obtains the sequence as 1, 2, 3, 4, 5, $\cdots$. Actually, the variable expression of this sequence is $a_n = 3n$. Since the first term is $a_1 = 3$, so, the expression $a_n = 3n$ is defined for $n \geq 1$.

Example 2. Write the general formula for the following sequence 5, 6, 7, 8, 9, $\cdots$ in terms of variable k.

Solution: See that, the common difference between the consecutive terms is $d = 1$. Subtract 4 from the terms of the sequence, then the obtained the sequence as 1, 2, 3, 4, 5, $\cdots$. Therefore, the general formula of this sequence in terms of variable k is $a_k = 4 + k$ such that $k \geq 1$.

Example 3. Find the variable expression corresponding to the following sequence.

$$3, 7, 11, 15, 19, \cdots$$

Solution: The given sequence is an arithmetic sequence, where each term increases by a common difference. We can see that the common difference (d) is: $7 - 3 = 4$. To find the general formula (or the nth term) of an arithmetic sequence, we use the formula: $x_n = x_1 + (n-1) \times d$, Where x_n is the nth term, x_1 is the first term, d is the common difference, and n is the position in the sequence. Given: $x_1 = 3$, and $d = 4$, it seems that the general rule is $x_n = 3 + 4(n-1) \rightarrow x_n = 3 + 4n - 4 \rightarrow x_n = 4n - 1$. Since the first term in the sequence is , the term x_n is defined for $n \geq 1$.

Write Variable Expressions for Geometric Sequences

- The given recursive sequence x_1, x_2, x_3, $\cdots$ such that the ratio of the consecutive terms is equal to the same value, which is called a geometric sequence.

- To write the variable expression for the geometric sequences, follow these steps:

  - Specify the initial values as $x_1 = a$.
  - Determine $r =$ the common ratio of the consecutive terms.
  - Multiply the first term by the product of the number of terms in the common ratio.

Examples:

Example 1. Write an equation to describe for following sequence 2, 4, 8, 16, 32, $\cdots$.

Solution: Let $x_1, x_2, x_3, \cdots$ equivalent to the sequence of the content question. So, the first term is $x_1 = 2$. Now, look for the common ratio between the consecutive terms. So, the common ratio is $r = 2$. Therefore, the sequence is geometric, and it can be rewritten as follows:

$$2,\ 2^2,\ 2^3,\ 2^4,\ 2^5,\ \cdots$$

You know that the first term of the sequence is multiple of 2, so, the variable expression of this sequence is $a_n = 2^n$. Since the first term is $a_1 = 2$, so, the obtained equation is defined for $n \geq 1$.

Example 2. Write the general formula for the following sequence 6, 18, 54, 162, 486, $\cdots$ in terms of variable i.

Solution: By dividing the consecutive terms by previous terms, you notice that the common ratio is $r = 3$. According to the terms of the sequence, you can see that all terms have the common multiple of 2. Divide all terms of the sequence by 2. So, the following sequence is obtained:

$$3,\ 9,\ 27,\ 81,\ 243,\ \cdots$$

It seems that the general formula of the last sequence in terms of variable i is $x_i = (3)^i$, where $i \geq 1$. Therefore, by multiplying this equation by 2, we have:

$$x_i = 2(3)^i, \text{ where } i \geq 1.$$

Write a Formula for a Recursive Sequence

- To find the general formula of a given recursive sequence, follow the steps below:
  - Specify the initial values as a_1 or even a_2.
  - Look for a relationship in the previous terms like a_{n-1} or even a_{n-2} that holds throughout the sequence. Start by evaluating the differences and ratios of consecutive terms.
  - Express the relationship as a function of the previous term or terms.

Examples:

Example 1. Find the recursive formula corresponding to the following sequence.

$$12, 17, 22, 27, 32, \cdots$$

Solution: Let the first term of this sequence $a_1 = 12$. To find the recursive formula, start by looking at the differences and ratios of consecutive terms. By evaluating the difference of terms with the previous term, you notice that the differences between consecutive terms are all the same. That is, $a_n - a_{n-1} = 5$. In this step, rewrite the obtained rule in terms of a_{n-1}. So, $a_n = a_{n-1} + 5$. Since the first term in the sequence is a_1, the term a_{n-1} is not defined for $n \leq 1$.

Example 2. Find the recursive formula for the following sequence.

$$1, 3, 9, 27, 81, \cdots$$

Solution: Here, let the first term of this sequence $a_1 = 1$. Looking at the terms in this sequence, you can see that each term has an equal ratio to the previous term. Thus, the following relationship is obtained: $\frac{a_n}{a_{n-1}} = 3$. Now, rewrite this rule in terms of a_{n-1}. Therefore, $a_n = 3a_{n-1}$. You know that $a_1 = 1$. It means that the terms a_{n-1} define for $n \geq 2$.

Example 3. Find the recursive formula corresponding to the following sequence.

$$1, 3, 2, -1, -3, -2, \cdots$$

Solution: Here, after evaluating the terms in the sequence, since the differences and ratios of consecutive terms are not the same value, therefore, look for another relationship between the previous term or even terms of the sequence. However, by evaluating the difference between the terms of the sequence, you can see that the difference of each term from the previous term makes the following sequence: $2, -1, -3, -2, +1 \cdots$. Actually, the term of the last sequence has the formula as $a_n - a_{n-1}$. Therefore, the obtained formula is the recursive

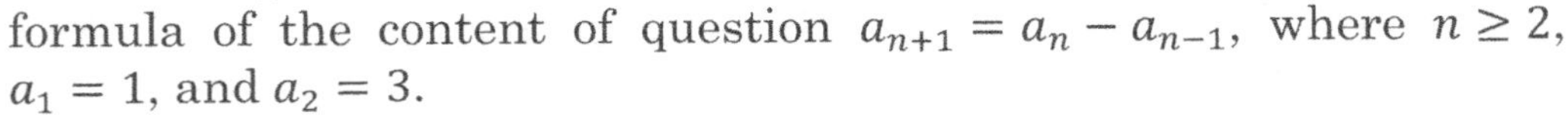

formula of the content of question $a_{n+1} = a_n - a_{n-1}$, where $n \geq 2$, $a_1 = 1$, and $a_2 = 3$.

Chapter 13: Practices

Solve.

1) Find the first four terms of the sequence with the general term $T_{n+1} = T_n + 5$, where $n \geq 1$ and $T_1 = -12$.
2) Find a_3, where the first term is $a_1 = 1$, and the general term is $a_n = 6a_{n-1}$.
3) Find a_6, where the first term is $a_1 = 3$, and the general term is $a_n = a_{n-1} + 8$.
4) Find the first four terms of the sequence with the general term $x_n = 2(3)^n$, where n represents the position of a term in the sequence and starts with $n = 1$.
5) Find the first three terms of the sequence with the general term $a_n = -7n - 2$, where n represents the position of a term in the sequence and starts with $n = 1$.
6) A sequence is defined by the formula $u_n = 3n + 1$, calculate the first 4 terms of this sequence.
7) Write the general formula for the following sequence 42, 84 , 126, 168, $\cdots$ in terms of variable k.
8) Find the variable expression corresponding to the following sequence. $-2, -6, -10, -14, -18, \ldots$
9) Write the general formula for the following sequence $-75, -74, -73, -72, \cdots$ in terms of variable n.
10) Write an equation to describe the following sequence. $1, -5, 25, \ldots.$
11) Write the general formula for the following sequence. $2, \frac{1}{2}, \frac{1}{8}, \frac{1}{32}, \frac{1}{128}, \ldots$
12) Find the general formula for the following sequence. $-2.5, -10, -40, -160, \ldots$
13) Find the recursive formula for the following arithmetic sequence: $1, 6, 11, 16, \ldots$
14) Write the recursive formula for the following sequence: $4, 11, 25, \ldots.$
15) Find the recursive formula corresponding to the following sequence. $-1, -4, -16, -64, \ldots$

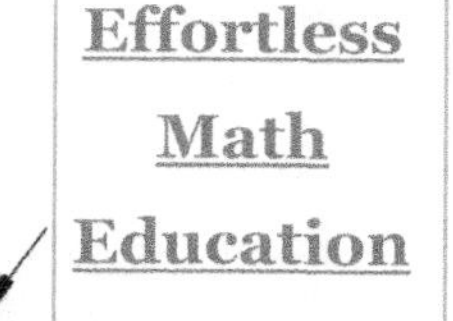

Chapter 13: Answers

1) $-12, -7, -2, 3$
2) 36
3) 43
4) $6, 18, 54, 162$
5) $-9, -16, -23$
6) $4, 7, 10, 13$
7) $a_k = 42k$
8) $a_n = -4n + 2$
9) $a_n = n - 76$
10) $a_n = 1(-5)^{n-1}$
11) $a_n = 2(\frac{1}{4})^{n-1}$
12) $a_n = -2.5(4)^{n-1}$
13) $a_n = a_{n-1} + 5, n \geq 2$
14) $a_{n+1} = a_n + 7n, n \geq 1$
15) $a_n = 4a_{n-1}, n \geq 2$

Time to Test

Time to refine your algebra skills with a practice test.

Take an Algebra I test to simulate the test day experience. After you've finished, score your test using the answer keys.

Before You Start

- You'll need a pencil and a calculator to take the test.
- For multiple questions, there are five possible answers. Choose which one is best.
- It's okay to guess. There is no penalty for wrong answers.
- Use the answer sheet provided to record your answers.
- **Scientific calculator is permitted for Algebra I Test.**
- After you've finished the test, review the answer key to see where you went wrong.

Good Luck!

Algebra I Practice Test 1

2023

Total number of questions: 60
Total time: No time limit
Calculator is permitted for Algebra I Test.

Algebra I Practice Test Answer Sheet

Remove (or photocopy) this answer sheet and use it to complete the practice test.

Algebra I Practice Test 1 Answer Sheet					
1	Ⓐ Ⓑ Ⓒ Ⓓ Ⓔ	21	Ⓐ Ⓑ Ⓒ Ⓓ Ⓔ	41	Ⓐ Ⓑ Ⓒ Ⓓ Ⓔ
2	Ⓐ Ⓑ Ⓒ Ⓓ Ⓔ	22	Ⓐ Ⓑ Ⓒ Ⓓ Ⓔ	42	Ⓐ Ⓑ Ⓒ Ⓓ Ⓔ
3	Ⓐ Ⓑ Ⓒ Ⓓ Ⓔ	23	Ⓐ Ⓑ Ⓒ Ⓓ Ⓔ	43	Ⓐ Ⓑ Ⓒ Ⓓ Ⓔ
4	Ⓐ Ⓑ Ⓒ Ⓓ Ⓔ	24	Ⓐ Ⓑ Ⓒ Ⓓ Ⓔ	44	Ⓐ Ⓑ Ⓒ Ⓓ Ⓔ
5	Ⓐ Ⓑ Ⓒ Ⓓ Ⓔ	25	Ⓐ Ⓑ Ⓒ Ⓓ Ⓔ	45	Ⓐ Ⓑ Ⓒ Ⓓ Ⓔ
6	Ⓐ Ⓑ Ⓒ Ⓓ Ⓔ	26	Ⓐ Ⓑ Ⓒ Ⓓ Ⓔ	46	Ⓐ Ⓑ Ⓒ Ⓓ Ⓔ
7	Ⓐ Ⓑ Ⓒ Ⓓ Ⓔ	27	Ⓐ Ⓑ Ⓒ Ⓓ Ⓔ	47	Ⓐ Ⓑ Ⓒ Ⓓ Ⓔ
8	Ⓐ Ⓑ Ⓒ Ⓓ Ⓔ	28	Ⓐ Ⓑ Ⓒ Ⓓ Ⓔ	48	Ⓐ Ⓑ Ⓒ Ⓓ Ⓔ
9	Ⓐ Ⓑ Ⓒ Ⓓ Ⓔ	29	Ⓐ Ⓑ Ⓒ Ⓓ Ⓔ	49	Ⓐ Ⓑ Ⓒ Ⓓ Ⓔ
10	Ⓐ Ⓑ Ⓒ Ⓓ Ⓔ	30	Ⓐ Ⓑ Ⓒ Ⓓ Ⓔ	50	Ⓐ Ⓑ Ⓒ Ⓓ Ⓔ
11	Ⓐ Ⓑ Ⓒ Ⓓ Ⓔ	31	Ⓐ Ⓑ Ⓒ Ⓓ Ⓔ	51	Ⓐ Ⓑ Ⓒ Ⓓ Ⓔ
12	Ⓐ Ⓑ Ⓒ Ⓓ Ⓔ	32	Ⓐ Ⓑ Ⓒ Ⓓ Ⓔ	52	Ⓐ Ⓑ Ⓒ Ⓓ Ⓔ
13	Ⓐ Ⓑ Ⓒ Ⓓ Ⓔ	33	Ⓐ Ⓑ Ⓒ Ⓓ Ⓔ	53	Ⓐ Ⓑ Ⓒ Ⓓ Ⓔ
14	Ⓐ Ⓑ Ⓒ Ⓓ Ⓔ	34	Ⓐ Ⓑ Ⓒ Ⓓ Ⓔ	54	Ⓐ Ⓑ Ⓒ Ⓓ Ⓔ
15	Ⓐ Ⓑ Ⓒ Ⓓ Ⓔ	35	Ⓐ Ⓑ Ⓒ Ⓓ Ⓔ	55	Ⓐ Ⓑ Ⓒ Ⓓ Ⓔ
16	Ⓐ Ⓑ Ⓒ Ⓓ Ⓔ	36	Ⓐ Ⓑ Ⓒ Ⓓ Ⓔ	56	Ⓐ Ⓑ Ⓒ Ⓓ Ⓔ
17	Ⓐ Ⓑ Ⓒ Ⓓ Ⓔ	37	Ⓐ Ⓑ Ⓒ Ⓓ Ⓔ	57	Ⓐ Ⓑ Ⓒ Ⓓ Ⓔ
18	Ⓐ Ⓑ Ⓒ Ⓓ Ⓔ	38	Ⓐ Ⓑ Ⓒ Ⓓ Ⓔ	58	Ⓐ Ⓑ Ⓒ Ⓓ Ⓔ
19	Ⓐ Ⓑ Ⓒ Ⓓ Ⓔ	39	Ⓐ Ⓑ Ⓒ Ⓓ Ⓔ	59	Ⓐ Ⓑ Ⓒ Ⓓ Ⓔ
20	Ⓐ Ⓑ Ⓒ Ⓓ Ⓔ	40	Ⓐ Ⓑ Ⓒ Ⓓ Ⓔ	60	Ⓐ Ⓑ Ⓒ Ⓓ Ⓔ

1) Which of the following points lies on the line $2x + 4y = 6$?

A. $(2,1)$

B. $(-1,2)$

C. $(-2,2)$

D. $(2,2)$

E. $(2,8)$

2) Point A lies on the line with equation $y - 3 = 2(x + 5)$. If the x −coordinate of A is 8, what is the y −coordinate of A?

A. 14

B. 16

C. 22

D. 29

E. 31

3) Write the equation of a horizontal line that goes through the point $(-5,12)$.

A. $y = 12$

B. $x = 12$

C. $y = -5$

D. $x = -5$

E. $x = 5$

4) The ratio of boys to girls in a school is $2:3$. If there are 500 students in a school, how many boys are in the school?

A. 540

B. 360

C. 300

D. 280

E. 200

5) Find the domain of the function represented in the graph.

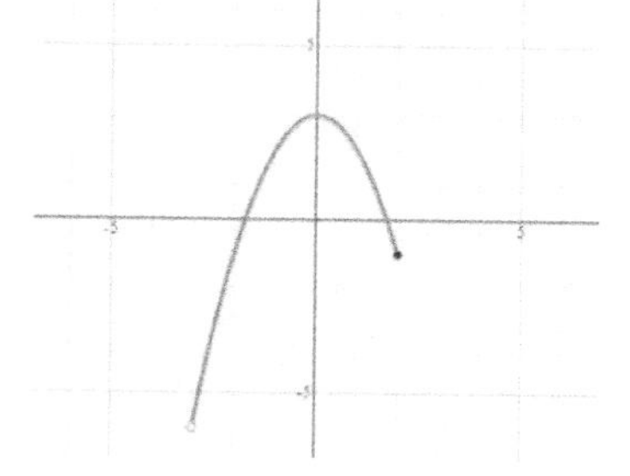

A. $(-3,2]$

B. $(-6,3)$

C. $(-3,2)$

D. $(-6,-1]$

E. $[-6,-1)$

6) $(7x+2y)(5x+2y)=?$

A. $2x^2+14xy+2y^2$

B. $2x^2+4xy+2y^2$

C. $7x^2+14xy+y^2$

D. $12x^2+14xy+4y$

E. $35x^2+24xy+4y^2$

7) Which of the following expressions is equivalent to $5x(4+2y)$?

A. $x+10xy$

B. $5x+5xy$

C. $20xy+2xy$

D. $20x+5xy$

E. $20x+10xy$

8) If $y=5ab+3b^3$, what is y when $a=2$ and $b=3$?

A. 24

B. 31

C. 36

D. 57

E. 111

9) Which of the following statement about the following graph is true?

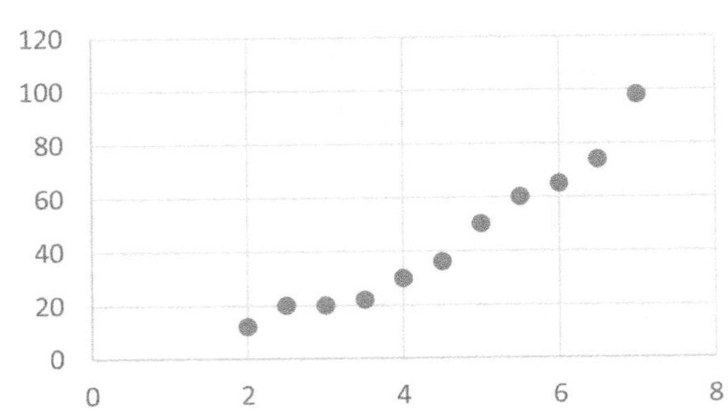

A. The scatter plot shows a negative trend.

B. The scatter plot shows a no trend.

C. The scatter plot shows a positive trend.

D. The scatter plot has outlier.

E. Choice C and D.

10) A bag contains 30 balls: ten green, nine black, eight blue, one brown, one red, and one white. If 29 balls are removed from the bag at random, what is the probability that a brown ball has been removed?

A. $\frac{1}{30}$

B. $\frac{3}{30}$

C. $\frac{27}{30}$

D. $\frac{29}{30}$

E. $\frac{29}{100}$

11) What is the solution to the following inequality?

$$|x - 10| \leq 3$$

A. $x \geq 13 \cup x \leq 7$

B. $7 \leq x \leq 13$

C. $-7 \leq x \leq 13$

D. $x \geq 13$

E. $x \leq 7$

12) Two third of 15 is equal to $\frac{2}{5}$ of what number?

A. 12

B. 20

C. 25

D. 60

E. 90

13) The marked price of a computer is D dollar. Its price decreased by 25% in January and later increased by 10% in February. What is the final price of the computer in D dollar?

A. $0.8D$

B. $0.825D$

C. $0.9D$

D. $1.2D$

E. $1.4D$

14) Which system of equations is represented by the following graph?

A. $\begin{array}{c} x - 2y = 4 \\ y - 3 = x \end{array}$

B. $\begin{array}{c} y^3 - 1 = x \\ 2y - x = 1 \end{array}$

C. $\begin{array}{c} x^2 - 1 = y \\ x + y = 1 \end{array}$

D. $\begin{array}{c} x - y = -5 \\ y - 1 = x^2 \end{array}$

E. $\begin{array}{c} x^2 = y + 1 \\ x - y = 1 \end{array}$

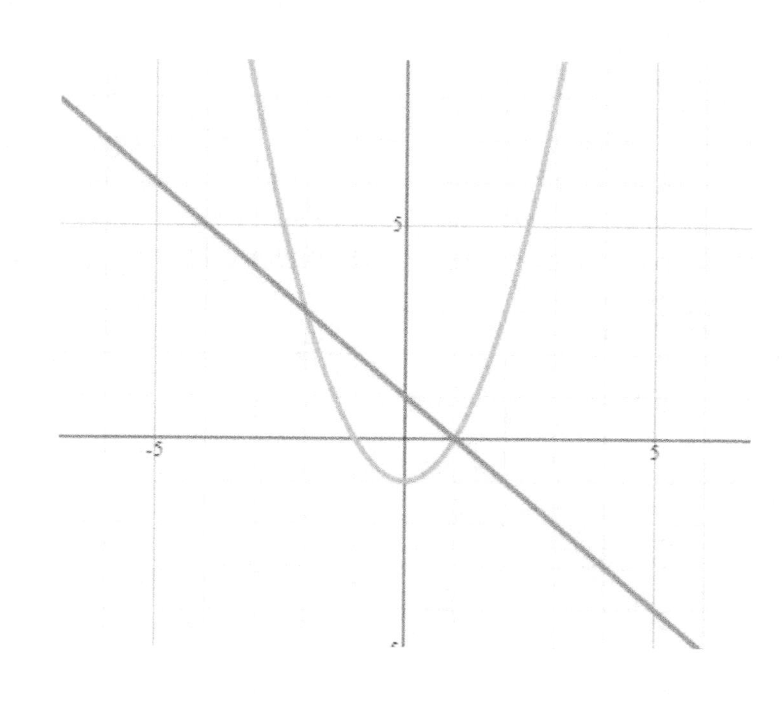

15) How many ways can we pick a team of 4 people from a group of 7?

A. 28

B. 30

C. 35

D. 40

E. 42

16) The average of 13, 15, 20 and x is 20. What is the value of x?

A. 9

B. 15

C. 18

D. 20

E. 32

17) The equation $x^2 = 4x - 3$ has how many distinct real solutions?

A. 0

B. 1

C. 2

D. 3

E. 4

18) What is the ratio of the minimum value to the maximum value of the following function?

$$f(x) = -3x + 1, -2 \leq x \leq 3$$

A. $\frac{6}{7}$

B. $\frac{7}{8}$

C. $-\frac{7}{8}$

D. $\frac{8}{7}$

E. $-\frac{8}{7}$

19) Simon has a total of 45 comic books and notebooks. If the number of comic books is five more than four times the number of notebooks, how many comic books does he have?

A. 32

B. 45

C. 37

D. 8

E. 40

20) For what value of x is $|x - 3| + 3$ equal to 0?

A. 2

B. -2

C. -3

D. 3

E. No values for x.

21) In 1999, the average worker's income increased by $2,000 per year starting from a $26,000 annual salary. Which equation represents income greater than average? (I = income, x = number of years after 1999)

A. $I > 2{,}000x + 26{,}000$

B. $I > -2{,}000x + 26{,}000$

C. $I < -2{,}000x + 26{,}000$

D. $I < 2{,}000x - 26{,}000$

E. $I < 24{,}000x + 26{,}000$

22) Solve: $\frac{3x+6}{x+5} \times \frac{x+5}{x+2} =$

A. 1

B. 2

C. 3

D. $\frac{x+5}{x+2}$

E. $3(x + 2)$

23) In the triangle below, if the measure of angle A is 37 degrees, then what is the value of y? (Figure is NOT drawn to scale)

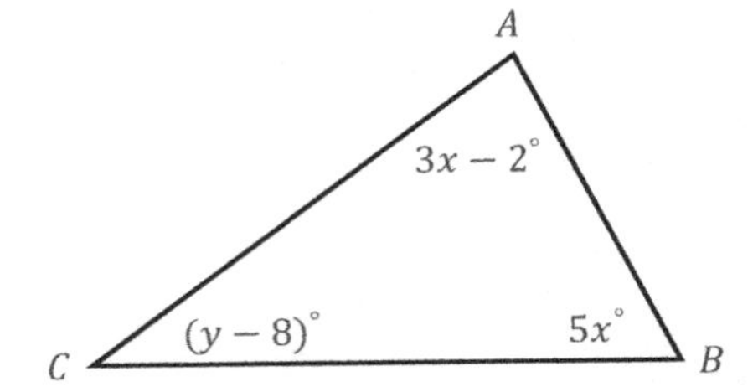

A. 70

B. 78

C. 84

D. 86

E. 92

24) The length of each side of a square box is represented by the expression $3\alpha^4$. The volume of the box is $(3\alpha^4)^3$.
Which simplified expression represents the volume of the box?

A. $27\alpha^7$

B. $9\alpha^8$

C. $27\alpha^{12}$

D. $9\alpha^{12}$

E. $3\alpha^{12}$

25) Which of the following is equivalent to $2\sqrt{32} + 2\sqrt{2}$?

A. $\sqrt{2}$

B. 2

C. $2\sqrt{2}$

D. $4\sqrt{2}$

E. $10\sqrt{2}$

26) Simplify. $\frac{6}{\sqrt{12}-3}$

A. $\sqrt{12} + 3$

B. 2

C. $2(\sqrt{12} + 3)$

D. $2\sqrt{12}$

E. 6

27) The average of five consecutive numbers is 36. What is the smallest number?

A. 38

B. 36

C. 34

D. 12

E. 8

28) Which of the following numbers is NOT a solution of the inequality $2x - 5 \geq 3x - 1$?

A. -2

B. -4

C. -5

D. -8

E. -10

29) If the following equations are true, what is the value of x?

$$a = \sqrt{3}, 4a = \sqrt{4x}$$

A. 2

B. 3

C. 6

D. 12

E. 14

30) In a greenhouse, the number of plants in each section depends on the number of sprinklers in that section. In a section with x sprinklers, there are $3x(x + 2)$ red rose plants and $(x + 3)(x + 1)$ white rose plants.
Which simplified expression represents the total number of red and white rose plants in a section with x sprinklers?

A. $3x^4 + 18x^3 + 33x^2 + 6x^3 + 18x$

B. $3x^4 + 18x^3 + 13x^2 + 6x^3 + 18x$

C. $3x^2 + 12x + 18$

D. $4x^2 + 10x + 3$

E. $6x^2 + 10x + 6$

31) If $\sqrt{4m-3} = m$, what is (are) the value(s) of m?

A. 0

B. 1

C. 1, 3

D. $-1, 3$

E. $-1, -3$

32) When a number is subtracted from 28 and the difference is divided by that number, the result is 3. What is the value of the number?

A. 2

B. 4

C. 7

D. 12

E. 24

33) An angle is equal to one-ninth of its supplement. What is the measure of that angle?

A. 9

B. 18

C. 25

D. 60

E. 90

34) What is the least common denominator for $\frac{3x}{x^2-36}$ and $\frac{4}{2x-12}$?

A. $2x^2 - 72$

B. $x^2 - 36$

C. $2x - 12$

D. $2x + 12$

E. $x^2 + 36$

35) Which of the following expressions is the inverse of the function? $h(x) = \sqrt{x} + 3$

A. $x^2 + 3$

B. $x^2 - 6x$

C. $x^2 - 6x + 12$

D. $x^2 - 6x + 9$

E. $x^2 - 12x + 9$

36) If $y = nx + 2$, where n is a constant, and when $x = 6$, $y = 14$, what is the value of y when $x = 10$?

A. 10

B. 12

C. 18

D. 22

E. 24

37) A chemical solution contains 6% alcohol. If there is 24 ml of alcohol, what is the volume of the solution?

A. 240 ml

B. 400 ml

C. 600 ml

D. 1,200 ml

E. 2,400 ml

38) The average weight of 18 girls in a class is 56 kg and the average weight of 32 boys in the same class is 62 kg. What is the average weight of all the 50 students in that class?

A. 50

B. 59.84

C. 61.68

D. 61.90

E. 62.20

39) The number of students on a basketball team and the number of free throws each student made during practice were put into a scatter plot.

Which is the outlier on the scatter plot? Specify the type of trend.

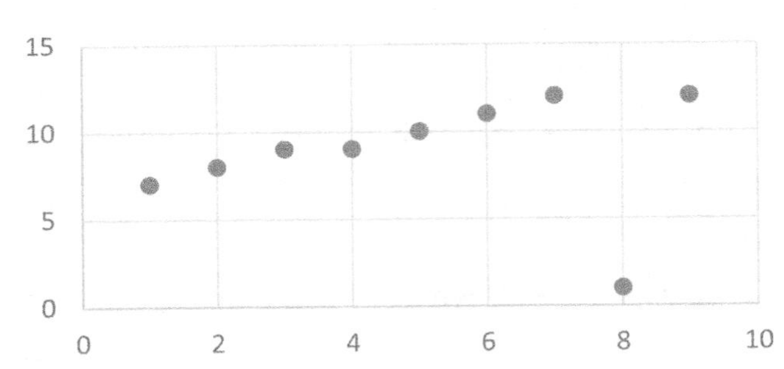

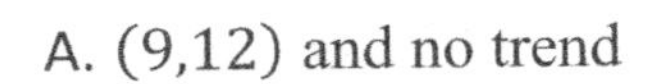
A. (9,12) and no trend

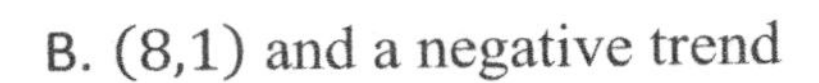
B. (8,1) and a negative trend

C. (9,12) and a positive trend

D. (8,1) and a positive trend

E. (8,1) and no trend

40) If 60% of x is equal to 30% of 20, then what is the value of $(x+3)^2$?

A. 25.25

B. 26

C. 26.01

D. 169

E. 225

41) If the function $g(x)$ has three distinct zeros, which of the following could represent the graph of $g(x)$?

A.

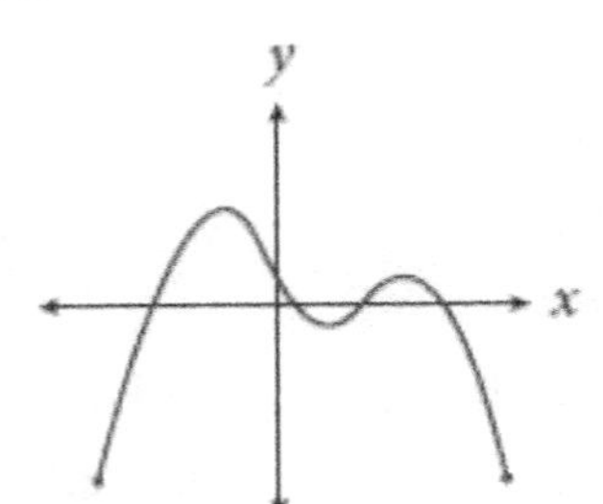

B.

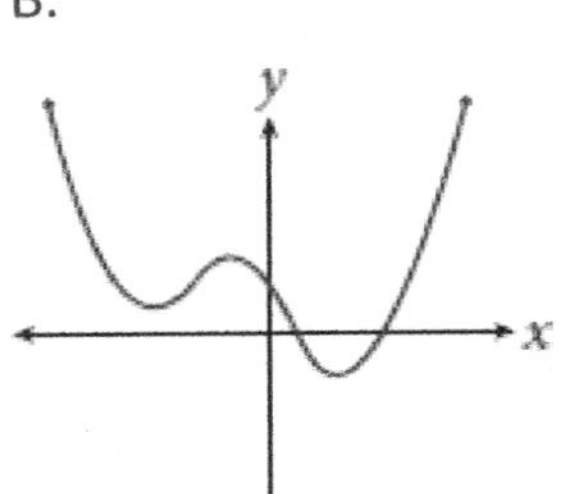

C.

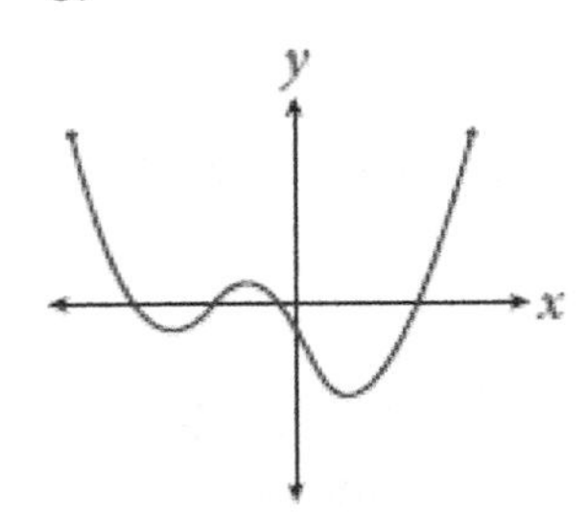

D.

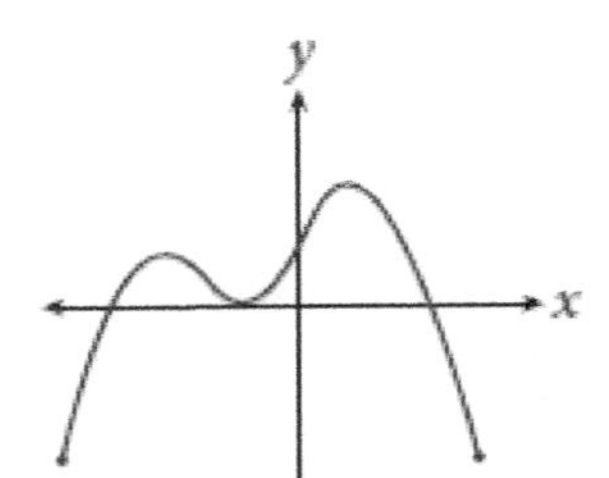

E.

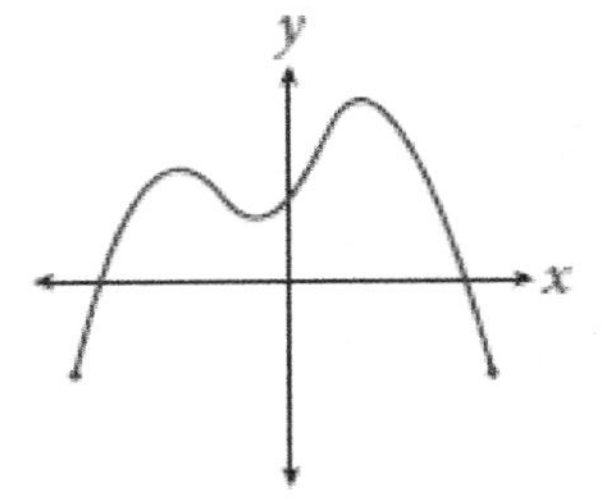

42) Multiply and write the product in scientific notation:

$$(2.9 \times 10^6) \times (2.6 \times 10^{-5})$$

A. 754×100

B. 75.4×10^6

C. 75.4×10^{-5}

D. 7.54×10^{11}

E. 7.54×10

43) 5 less than twice a positive integer is 73. What is the integer?

A. 39

B. 41

C. 42

D. 44

E. 50

44) The circle graph below shows all Mr. Green's expenses for last month. If he spent $660 on his car, how much did he spend for his rent?

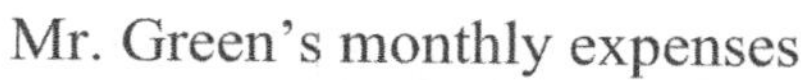

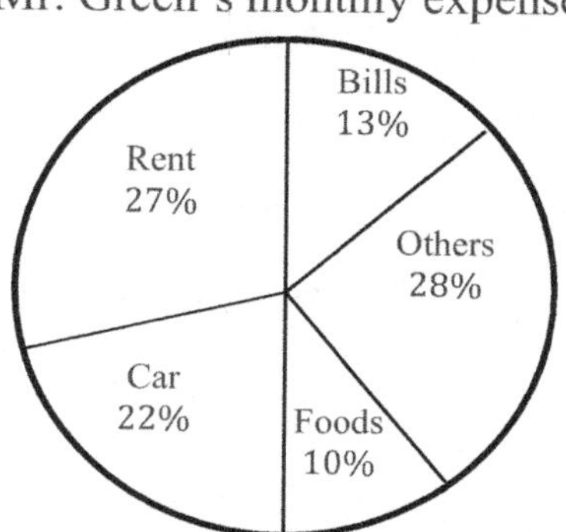

A. $700

B. $740

C. $810

D. $910

E. $660

45) If $f(x) = 3x + 4(x + 1) + 2$ then $f(4x) =?$

A. $28x + 6$

B. $25x + 4$

C. $16x - 6$

D. $12x + 3$

E. $12x - 3$

46) If $2x + 2y = 2$, $3x - y = 7$, which of the following ordered pairs (x, y) satisfies both equations?

A. $(1,3)$

B. $(2,4)$

C. $(2,-1)$

D. $(4,-6)$

E. $(1,-6)$

47) A line in the xy −plane passes through the origin and has a slope of $\frac{1}{3}$. Which of the following points lies on the line?

A. $(2,1)$

B. $(4,1)$

C. $(9,3)$

D. $(6,3)$

E. $(1,3)$

48) Which of the following is equivalent to $(3n^2 + 2n + 6) - (2n^2 - 4)$?

A. $n + 2$

B. $n - 2$

C. $n^2 - 3$

D. $n^2 + 2n - 2$

E. $n^2 + 2n + 10$

49) Solve for x: $4(x + 1) = 6(x - 4) + 20$.

A. 0

B. 2

C. 4

D. 6.5

E. 12

50) What is the value of x in the following equation? $3x + 10 = 46$

A. 4

B. 7

C. 10

D. 12

E. 16

51) If $x \neq -4$ and $x \neq 5$, which of the following is equivalent to $\frac{1}{\frac{1}{x-5}+\frac{1}{x+4}}$?

A. $\frac{(x-5)(x+4)}{(x-5)+(x+4)}$

B. $\frac{(x+4)+(x-5)}{(x+4)(x-5)}$

C. $\frac{(x+4)(x-5)}{(x+4)-(x+5)}$

D. $\frac{(x+4)+(x-5)}{(x+4)-(x-5)}$

E. $\frac{(x-4)+(x-5)}{(x+4)-(x-5)}$

$$y < c - x, y > x + b$$

52) In the xy −plane, if (0,0) is a solution to the system of inequalities above, which of the following relationships between c and b must be true?

A. $c < b$

B. $c > b$

C. $c = b$

D. $c = b + c$

E. $c = b - x$

53) What is the value of x in this equation?

$$4\sqrt{2x+6} = 24$$

A. 6

B. 36

C. 30

D. 15

E. 24

54) Calculate $f(5)$ for the following function f.

$$f(x) = x^2 - 3x$$

A. 5

B. 10

C. 15

D. 20

E. 25

55) John buys a pepper plant that is 5 inches tall. With regular watering, the plant grows 3 inches a year. Writing John's plant's height as a function of time, what does the $y-$intercept represent?

A. The $y-$intercept represents the rate of growth of the plant which is 5 inches.
B. The $y-$intercept represents the starting height of 5 inches.
C. The $y-$intercept represents the rate of growth of the plant which is 3 inches per year.
D. The $y-$intercept is always zero.
E. There is no $y-$intercept.

56) If $\frac{4}{x} = \frac{12}{x-8}$ what is the value of $\frac{x}{2}$?

A. 0

B. 1

C. 3

D. -2

E. 2

57) The Jackson Library is ordering some bookshelves. If x is the number of bookshelves the library wants to order, which each costs $\$200$ and there is a one-time delivery charge of $\$600$, which of the following represents the total cost, in dollars, per bookshelf?

A. $\frac{200x+600}{x}$

B. $\frac{200x+600}{200}$

C. $600x + 600$

D. $200 + 600x$

E. $200x + 600$

58) A function $g(3) = 5$ and $g(5) = 4$. A function $f(5) = 2$ and $f(4) = 6$. What is the value of $f(g(5))$?

A. 5

B. 6

C. 7

D. 8

E. 10

59) If $x \blacksquare y = \sqrt{x^2 + y}$, what is the value of $6 \blacksquare 28$?

A. 4

B. 6

C. 8

D. 10

E. $\sqrt{168}$

60) What is the value of y in the following system of equation?

$$3x - 4y = -20$$
$$-x + 2y = 10$$

A. 1

B. -1

C. -2

D. 4

E. 5

End of Algebra I Practice Test 1

Algebra I Practice Test 2

2023

Total number of questions: 60
Total time: No time limit
Calculator is permitted for Algebra I Test.

Algebra I Practice Test Answer Sheet

Remove (or photocopy) this answer sheet and use it to complete the practice test.

Algebra I Practice Test 2 Answer Sheet

#	Choices	#	Choices	#	Choices
1	A B C D E	21	A B C D E	41	A B C D E
2	A B C D E	22	A B C D E	42	A B C D E
3	A B C D E	23	A B C D E	43	A B C D E
4	A B C D E	24	A B C D E	44	A B C D E
5	A B C D E	25	A B C D E	45	A B C D E
6	A B C D E	26	A B C D E	46	A B C D E
7	A B C D E	27	A B C D E	47	A B C D E
8	A B C D E	28	A B C D E	48	A B C D E
9	A B C D E	29	A B C D E	49	A B C D E
10	A B C D E	30	A B C D E	50	A B C D E
11	A B C D E	31	A B C D E	51	A B C D E
12	A B C D E	32	A B C D E	52	A B C D E
13	A B C D E	33	A B C D E	53	A B C D E
14	A B C D E	34	A B C D E	54	A B C D E
15	A B C D E	35	A B C D E	55	A B C D E
16	A B C D E	36	A B C D E	56	A B C D E
17	A B C D E	37	A B C D E	57	A B C D E
18	A B C D E	38	A B C D E	58	A B C D E
19	A B C D E	39	A B C D E	59	A B C D E
20	A B C D E	40	A B C D E	60	A B C D E

1) If $f(x) = 2x + 2$ and $g(x) = x^2 + 4x$, then find $(\frac{f}{g})(x)$.

A. $\frac{2x+2}{x^2+4x}$

B. $\frac{x+1}{x^2+2x}$

C. $\frac{2x+2}{x^2+1}$

D. $\frac{2x+2}{x^2+x}$

E. $\frac{x^2+4x}{2x+2}$

2) In the standard (x, y) coordinate plane, which of the following lines contains the points $(3, -5)$ and $(8,15)$?

A. $y = \frac{1}{4}x + 13$

B. $y = -\frac{1}{4}x + 17$

C. $y = 2x - 11$

D. $y = 4x - 17$

E. $y = -4x + 7$

3) Which of the following is equal to the expression below?

$$(5x + 2y)(2x - y)$$

A. $2x^2 + 6xy - 2y^2$

B. $4x^2 - 2y^2$

C. $8x^2 + 2xy - 2y^2$

D. $10x^2 - xy - 2y^2$

E. $24x^2 + 2xy - 2y^2$

4) What is the product of all possible values of x in the following equation? $|x-10|=4$

A. 3

B. 7

C. 13

D. 84

E. 100

5) How many vertical asymptotes does the graph of $y=\frac{2x-1}{x^2+2}$ have?

A. 0 vertical asymptotes

B. 1 vertical asymptotes

C. 2 vertical asymptotes

D. 3 vertical asymptotes

E. 4 vertical asymptotes

6) What is the slope of a line that is perpendicular to the line $4x-2y=6$?

A. $-\frac{1}{2}$

B. -2

C. 4

D. 12

E. 14

7) What is the value of the expression $6(x-2y)+(2-x)^2$ when $x=3$ and $y=-2$?

A. -4

B. 20

C. 43

D. 50

E. 80

8) Jack made a scatter plot of how much money he had left at the end of each week of the month.

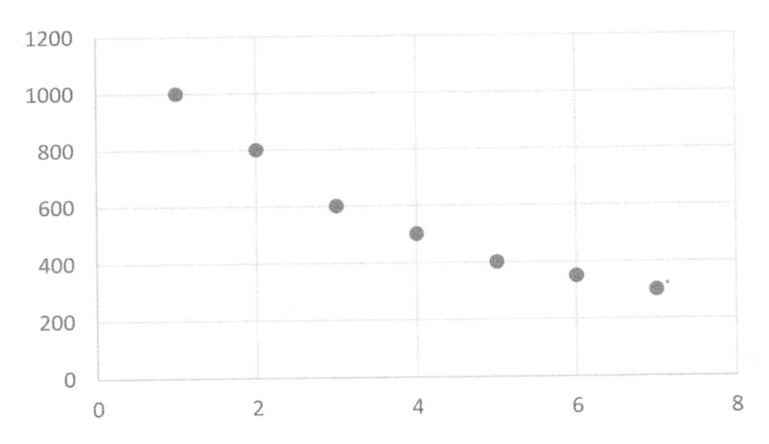

Which table best represents the data in her scatter plot?

A.

Day	1	2	3	4	5	6	7
Money	\$300	\$350	\$400	\$500	\$600	\$800	\$1000

B.

Day	1	2	3	4	5	6	7
Money	\$1000	\$1000	\$600	\$500	\$400	\$350	\$300

C.

Day	1	2	3	4	5	6	7
Money	\$100	\$800	\$600	\$500	\$400	\$350	\$300

D.

Day	1	2	3	4	5	6	7
Money	\$1000	\$800	\$600	\$500	\$400	\$350	\$300

E.

Day	1	2	3	4	5	6	7
Money	\$1000	\$800	\$700	\$500	\$400	\$350	\$100

9) If a function is defined as $f(x) = bx^2 + 15$, b is a constant and $f(2) = 35$. What is the value of $f(3)$?

A. 25

B. 35

C. 60

D. 65

E. 75

10) What is the vertical asymptote of the graph of $f(x) = \frac{1}{x^3+8}$?

A. $x = 2$

B. $x = -2$

C. $x = -8$

D. $y = -2\sqrt{2}$

E. $y = -x$

11) Find the domain of the radical following function. $y = 7\sqrt{5x - 15} + 5$

A. $x \geq 0$

B. $x \geq 1$

C. $x \geq 2$

D. $x \geq 3$

E. $x \geq 5$

12) The average of five numbers is 26. If a sixth number 42 is added, then, what is the new average? (Round your answer to the nearest hundredth.)

A. 25

B. 26.51

C. 27

D. 28.67

E. 36

13) A construction company is building a wall. The company can build 30 cm of the wall per minute. After 40 minutes $\frac{3}{4}$ of the wall is completed. How many meters is the wall?

A. 6

B. 8

C. 14

D. 16

E. 20

14) What is the solution to the following inequality?

$$|x-2| \geq 3$$

A. $x \leq -1 \cup x \geq 5$

B. $-1 \leq x \leq 5$

C. $1 \leq x \leq 5$

D. $x \geq 5$

E. $x \leq -1$

15) The graph of $f(x) = -x^2$ was translated 2 units to the right to create the graph of function g. Which function represents g?

A. $g(x) = -(x+2)^2$

B. $g(x) = (-x+2)^2$

C. $g(x) = -(x-2)^2$

D. $g(x) = -x^2 + 2$

E. $g(x) = x^2 + 2$

16) When 5 times the number x is added to 10, the result is 35. What is the result when 3 times x is added to 6?

A. 10

B. 15

C. 21

D. 25

E. 28

17) If $3h + g = 8h + 4$, what is g in terms of h?

A. $h = 5g - 4$

B. $g = 5h + 4$

C. $h = 4g$

D. $g = h + 1$

E. $g = 5h + 1$

18) What is the value of x in the following equation? $\frac{2}{3}x + \frac{1}{6} = \frac{1}{2}$

A. 2

B. $\frac{1}{2}$

C. $\frac{1}{3}$

D. $\frac{1}{4}$

E. $\frac{1}{12}$

19) A bank is offering 4.5% simple interest on a savings account. If you deposit $12,000, how much interest will you earn in two years?

A. $540

B. $1,080

C. $4,200

D. $8,400

E. $9,600

20) How many ways can a group of 4 be chosen from of 12 people?

A. 64

B. 120

C. 294

D. 495

E. 184

21) Simplify. $7x^2y^3(2x^2y)^3 =$

A. $14x^4y^6$

B. $14x^8y^6$

C. $56x^4y^6$

D. $56x^8y^6$

E. $96x^8y^6$

22) What are the zeroes of the function $f(x) = x^3 + 7x^2 + 12x$?

A. 0

B. $-4, -3$

C. $0, 2, 3$

D. $-3, -5$

E. $0, -3, -4$

23) Simplify. $-15a(a + b)^2 + 23a(a + b)^2 =$

A. 8

B. $8a$

C. $8(a + b)$

D. $8(a + b)^2$

E. $8a(a + b)^2$

24) If $\sqrt{3x} = \sqrt{y}$, then $x =$?

A. $3y$

B. $\sqrt{\frac{y}{3}}$

C. $\sqrt{3y}$

D. y^2

E. $\frac{y}{3}$

25) Solve for the inequality of $8x - 4 \geq -2x + 16$.

A. $x < 2$

B. $x \leq 2$

C. $x \geq 2$

D. $x > 2$

E. $x \geq -2$

26) If $f(x)=2x^3+5x^2+2x$ and $g(x)=-4$, what is the value of $f(g(x))$?

A. -4

B. 24

C. 32

D. 56

E. -56

27) What is the equivalent temperature of $104^{\circ}F$ in Celsius?

$$C=\frac{5}{9}(F-32)$$

A. 32

B. 40

C. 48

D. 52

E. 64

28) The average of 6 numbers is 14. The average of 4 of those numbers is 10. What is the average of the other two numbers?

A. 10

B. 12

C. 14

D. 22

E. 24

29) If 150% of a number is 75, then what is the 80% of that number?

A. 40

B. 50

C. 70

D. 85

E. 90

30) Find the domain and the range of function: $f(x) = \frac{1}{x+2}$.

A. The domain $\mathbb{R} - \{-2\}$ and range $\mathbb{R} - \{0\}$,

B. The domain $\mathbb{R} - \{0\}$ and range $\mathbb{R} - \{-2\}$,

C. The domain $\mathbb{R}$ and range $\mathbb{R} - \{0\}$,

D. The domain $\mathbb{R} - \{-2\}$ and range $\mathbb{R}$,

E. The domain $\mathbb{R} - \{-2,0\}$ and range $\mathbb{R} - \{0\}$.

31) What is the slope of the line? $4x - 2y = 12$

A. 1

B. -1

C. 1.5

D. -2

E. 2

32) In two successive years, the population of a town is increased by 10% and 20%. What percent of the population is increased after two years?

A. 30%

B. 32%

C. 35%

D. 68%

E. 70%

33) If 20% of a number is 4, what is the number?

A. 4

B. 8

C. 10

D. 20

E. 25

34) What is the value of x in the following system of equations?

$$5x + 2y = 3$$
$$y = x$$

A. $x = \frac{1}{3}$

B. $x = \frac{2}{3}$

C. $x = \frac{3}{7}$

D. $x = \frac{4}{3}$

E. $x = \frac{5}{3}$

35) Which multiplied expression represents the $3x^4u^5 \times 5u^2 \times 2x$?

A. $30x^7u^5$

B. $10x^5u^7$

C. $10x^7u^5$

D. $30x^5u^7$

E. $30x^7u^7$

36) In a hotel, there are 5 floors and x rooms on each floor. If each room has exactly y chairs, which of the following gives the total number of chairs in the hotel?

A. $2xy$

B. $5xy$

C. $x + y$

D. $x + 5y$

E. $2x + 5y$

37) If $\alpha = 2\beta$ and $\beta = 3\gamma$, how many α are equal to 36γ?

A. 1

B. 2

C. 4

D. 6

E. 12

38) If $f(x)=2x^3+2$ and $g(x)=\frac{1}{x}$, what is the value of $f(g(x))$?

A. $\frac{1}{2x}$

B. $\frac{2}{x^3}$

C. $\frac{1}{2x+2}$

D. $\frac{1}{2x^3+2}$

E. $\frac{2}{x^3}+2$

39) What are the coordinates of the x −intercept in the graph below?

A. $(0,3)$

B. $(0,2)$

C. $(-2,3)$

D. $(-3,0)$

E. $(-3,2)$

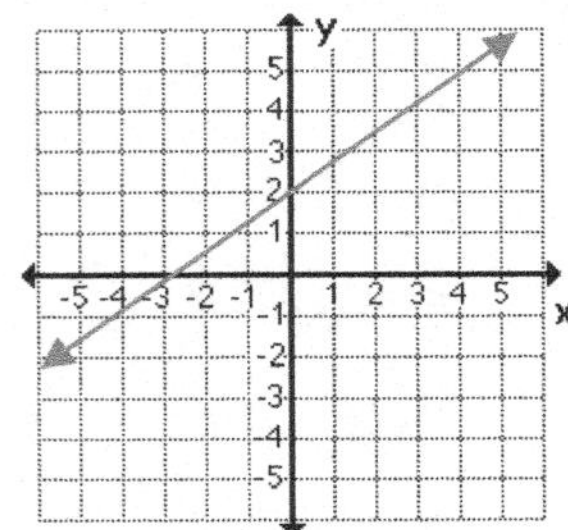

40) Which simplified expression represents the equation $\sqrt{\frac{x^2}{2}+\frac{x^2}{16}}$?

A. $\frac{4}{3}x$

B. $\frac{3}{4}x$

C. $3x$

D. $\frac{9}{16}x$

E. $9x$

41) If the ratio of $5a$ to $2b$ is $\frac{1}{10}$, what is the ratio of a to b?

A. $\frac{1}{10}$

B. $\frac{1}{20}$

C. $\frac{1}{25}$

D. 10

E. 25

42) If $x = 9$, what is the value of y in the following equation? $2y = \frac{2x^2}{3} + 6$

A. 30

B. 45

C. 60

D. 120

E. 180

43) If $\frac{x-3}{5} = N$ and $N = 6$, what is the value of x?

A. 25

B. 28

C. 30

D. 33

E. 36

44) Which of the following is equal to $b^{\frac{3}{5}}$?

A. $b^{\frac{5}{3}}$

B. $\sqrt{b^{\frac{5}{3}}}$

C. $\sqrt[5]{b^3}$

D. $\sqrt[3]{b^5}$

E. $\sqrt[3]{b^{-5}}$

45) On Saturday, Sara read N pages of a book each hour for 3 hours, and Mary read M pages of a book each hour for 4 hours. Which of the following represents the total number of pages of the book read by Sara and Mary on Saturday?

A. $7MN$

B. $12MN$

C. $3N + 4M$

D. $4N + 3M$

E. $4N - 3M$

46) Which set of data best represents the data on the scatter plot?

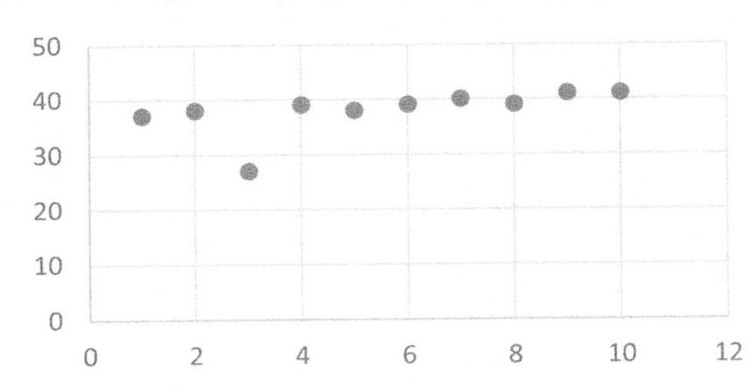

A.

Student	1	2	3	4	5	6	7	8	9	10
Size	41	40	39	40	39	38	39	27	38	37

B.

Student	1	2	3	4	5	6	7	8	9	10
Size	37	38	27	39	38	39	48	39	40	26

C.

Student	1	2	3	4	5	6	7	8	9	10
Size	37	38	27	39	38	39	40	39	41	41

D.

Student	37	38	27	39	38	39	40	39	40	41
Size	1	2	3	4	5	6	7	8	9	10

E.

Student	1	2	3	4	5	6	7	8	9	10
Size	41	38	44	39	38	39	41	41	41	41

47) Which of the following represents the graph of the line with the following equation?

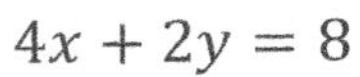

$$4x + 2y = 8$$

A.

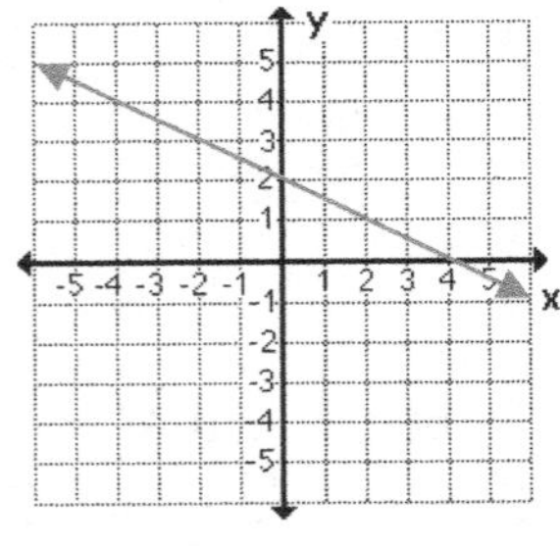

B.

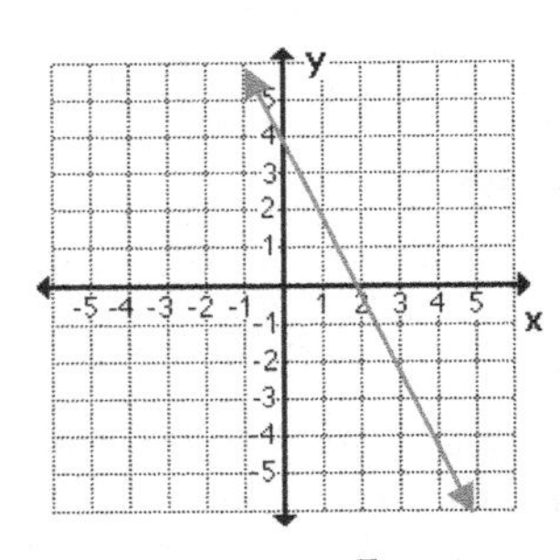

C.

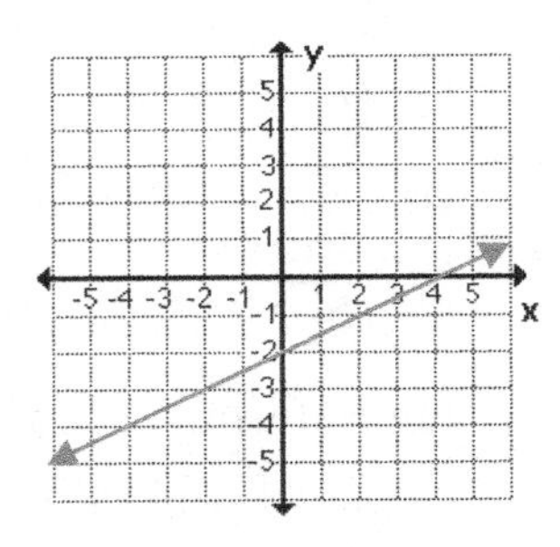

D.

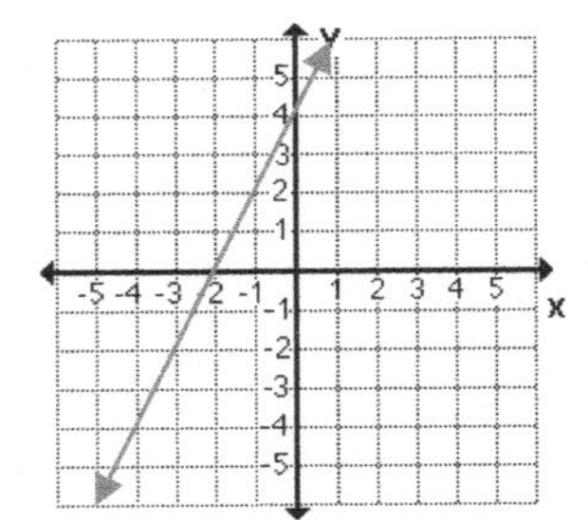

E.

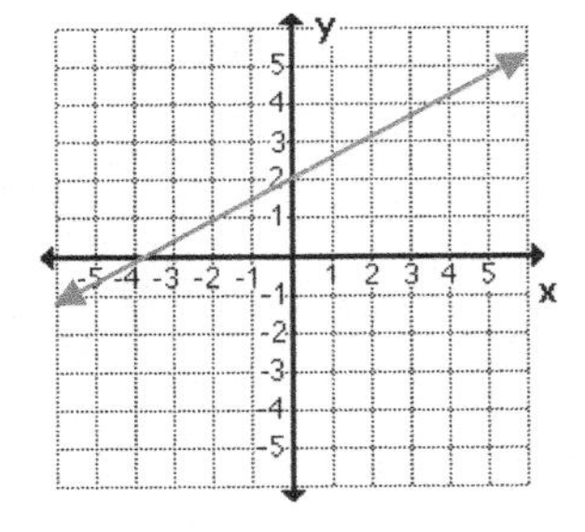

48) Find the solution (x, y) to the following system of equations.

$$2x + 5y = 11$$
$$4x - 2y = -14$$

A. $(2,3)$

B. $(-2,3)$

C. $(6,8)$

D. $(11,17)$

E. $(14,5)$

49) Which of the following is the largest?

A. $|4 - 2|$

B. $|2 - 4|$

C. $|-2 - 4|$

D. $|2 - 4| - |4 - 2|$

E. $|2 - 4| + |4 - 2|$

50) In the following figure, $ABCD$ is a rectangle. If $a = \sqrt{3}$, and $b = 2a$, find the area of the shaded region. (The shaded region is a trapezoid.)

A. $3\sqrt{3}$

B. $4\sqrt{3}$

C. 12

D. 3

E. $12\sqrt{3}$

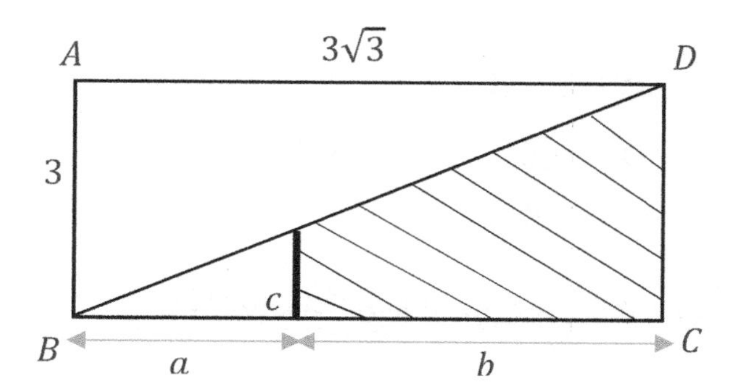

51) Which equation most closely models the data in the scatter plots?

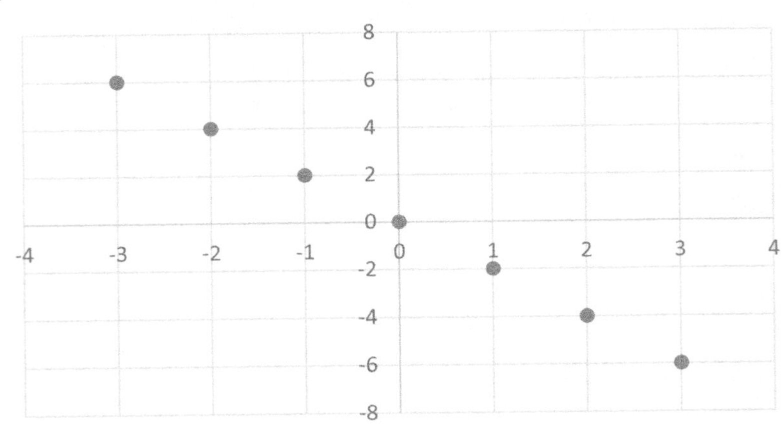

A. $y = x$

B. $y = -x$

C. $y = x + 1$

D. $y = -2x$

E. $y = 2x$

52) The function $g(x)$ is defined by a polynomial. Some values of x and $g(x)$ are shown in the table below. Which of the following must be a factor of $g(x)$?

A. x

B. $x - 1$

C. $x + 1$

D. $x - 2$

E. $x + 6$

x	$g(x)$
0	5
1	4
2	0

53) What is the value of $\frac{4b}{c}$ when $\frac{c}{b} = 2$?

A. 0

B. 1

C. 2

D. 4

E. 8

54) If $x + 5 = 8$, $2y - 1 = 5$ then $xy + 15 = ?$

A. 30

B. 24

C. 21

D. 17

E. 15

55) Which point lies in the solution set for the following system of inequalities?

$$y < \frac{1}{5}x + 3$$
$$2y + x \geq 1$$

A. $(-1,0)$

B. $(0,4)$

C. $(1,4)$

D. $(1,0)$

E. $(0.-1)$

56) If $\frac{a-b}{b} = \frac{10}{13}$, then which of the following must be true?

A. $\frac{a}{b} = \frac{10}{13}$

B. $\frac{a}{b} = \frac{10}{23}$

C. $\frac{a}{b} = \frac{13}{21}$

D. $\frac{a}{b} = \frac{21}{10}$

E. $\frac{a}{b} = \frac{23}{13}$

57) Which of the following lines is parallel to: $6y - 2x = 24$?

A. $y = x - 2$

B. $y = -x - 1$

C. $y = \frac{1}{3}x + 2$

D. $y = 2x - 1$

E. $y = 3x + 5$

58) 12 Students had a mean score of 75. The remaining 10 students in the class had an average score of 85. What is approximately the mean (average) score of the entire class?

A. 73.5

B. 75.5

C. 77.5

D. 79.5

E. 81.5

59) Sara orders a box of pens for $3 per box. A tax of 8.5% is added to the cost of the pens before a flat shipping fee of $6 closest out the transaction. Which of the following represents the total cost of p boxes of pens in dollars?

A. $3p + 6$

B. $6p + 3$

C. $6p + 6$

D. $1.085(6p) + 3$

E. $1.085(3p) + 6$

60) A plant grows at a linear rate. After five weeks, the plant is 40 cm tall. Which of the following functions represents the relationship between the height (y) of the plant and the number of weeks of growth (x)?

A. $y(x) = 40x + 8$

B. $y(x) = 8x + 40$

C. $y(x) = 40x$

D. $y(x) = 8x$

E. $y(x) = 4x$

End of Algebra I Practice Test 2

Algebra I Practice Tests Answer Keys

Now, it's time to review your results to see where you went wrong and what areas you need to improve.

Algebra I Practice Test 1						Algebra I Practice Test 2					
1	B	21	A	41	D	1	A	21	D	41	C
2	D	22	C	42	E	2	D	22	E	42	A
3	A	23	D	43	A	3	D	23	E	43	D
4	E	24	C	44	C	4	D	24	E	44	C
5	A	25	E	45	A	5	A	25	C	45	C
6	E	26	C	46	C	6	A	26	E	46	C
7	E	27	C	47	C	7	C	27	B	47	B
8	E	28	A	48	E	8	D	28	D	48	B
9	C	29	D	49	C	9	C	29	A	49	C
10	D	30	D	50	D	10	B	30	A	50	B
11	B	31	C	51	A	11	D	31	E	51	D
12	C	32	C	52	B	12	D	32	B	52	D
13	B	33	B	53	D	13	D	33	D	53	C
14	C	34	A	54	B	14	A	34	C	54	B
15	C	35	D	55	B	15	C	35	D	55	D
16	E	36	D	56	D	16	C	36	B	56	E
17	C	37	B	57	A	17	B	37	D	57	C
18	E	38	B	58	B	18	B	38	E	58	D
19	C	39	D	59	C	19	B	39	D	59	E
20	E	40	D	60	E	20	D	40	B	60	D

Algebra I Practice Test 1 Answers and Explanations

Algebra I Practice Tests 1 Explanations

1) Choice B is correct

Plug in each pair of number in the equation:

A. (2,1): $2(2) + 4(1) = 8$ This is NOT true.

B. (−1,2): $2(-1) + 4(2) = 6$ This is true.

C. (−2,2): $2(-2) + 4(2) = 4$ This is NOT true.

D. (2,2): $2(2) + 4(2) = 12$ This is NOT true.

E. (2,8): $2(2) + 4(8) = 36$ This is NOT true.

2) Choice D is correct

Here we can substitute 8 for x in the equation. Thus, $y - 3 = 2(8 + 5)$, $y - 3 = 26$.

Adding 3 to both sides of the equation: $y = 26 + 3$, $y = 29$.

3) Choice A is correct

Since the line is horizontal, the equation of the line is in the form of: $y = b$, where y always takes the same value of 12. Thus, the equation of the line is: $y = 12$

4) Choice E is correct

The ratio of boys to girls is $2:3$. Therefore, there are 2 boys out of 5 students. To find the answer, first divide the total number of students by 5, then multiply the result by 2.

$$500 \div 5 = 100 \rightarrow 100 \times 2 = 200$$

5) Choice A is correct

The domain is the set of x (or input) values that will satisfy the function. In this graph, we have a curve from the point $(-3, -6)$ to $(2, -1)$, where $(-3, -6)$ is not on the graph. Therefore the x (or input values) range from $(-3,2]$.

6) Choice E is correct

Use FOIL (First, Out, In, Last) method:

$$(7x + 2y)(5x + 2y) = 35x^2 + 14xy + 10xy + 4y^2 = 35x^2 + 24xy + 4y^2.$$

7) Choice E is correct

Use distributive property: $5x(4 + 2y) = 20x + 10xy$.

8) Choice E is correct

$y = 5ab + 3b^3$. Plug in the values of a and b in the equation: $a = 2$ and $b = 3$.

$$y = 5(2)(3) + 3(3)^3 = 30 + 3(27) = 30 + 81 = 111.$$

9) Choice C is correct

We know that a scatter plot shows a positive trend if y tends to increase as x increases. Taking that into consideration when looking at the scatter plot, it is noticeable that the trend is positive.

10) Choice D is correct

If 29 balls are removed from the bag at random, there will be one ball in the bag. The probability of choosing a brown ball is 1 out of 30. Therefore, the probability of not choosing a brown ball is 29 out of 30 and the probability of having not a brown ball after removing 29 balls is the same. The answer is: $\frac{29}{30}$.

11) Choice B is correct

$$|x - 10| \le 3 \rightarrow -3 \le x - 10 \le 3 \rightarrow -3 + 10 \le x - 10 + 10 \le 3 + 10 \rightarrow 7 \le x \le 13$$

12) Choice C is correct

Let x be the number. Write the equation and solve for x. $\frac{2}{3} \times 15 = \frac{2}{5} . x \rightarrow \frac{2\times15}{3} = \frac{2x}{5}$, use cross multiplication to solve for x. $5 \times 30 = 2x \times 3 \rightarrow 150 = 6x \rightarrow x = 25$

13) Choice B is correct

To find the discount, multiply the number by (100% *−rate of discount*). Therefore, for the first discount we get: $(D)(100\% - 25\%) = (D)(0.75) = 0.75D$. For increase of 10%:

$$(0.75D)(100\% + 10\%) = (0.75D)(1.10) = 0.825D$$

14) Choice C is correct

This question may be solved by either using the graph to determine the equation of each line or using the solution to the system of equations, $(-2,3)$ and $(1,0)$.

Method 1: Using the graph to determine the equation of each line:

Clearly, a red line is a linear equation that passes through two points $(-2,3)$ and $(1,0)$. Then, the general slope-intercept form of the equation of a line is $y = mx + b$, where m is

the slope and b is the y −intercept. Now, find the slope: $m = \frac{y_2 - y_1}{x_2 - x_1} = \frac{3-0}{-2-1} = \frac{3}{-3} = -1$.
Now, we have: $y = -x + b$. To find the value of b:

$$(1,0) \rightarrow y = -x + b \rightarrow 0 = -1 + b \rightarrow b = 1.$$

Therefore: $y = -x + 1$.

The blue line is a quadratic function with the vertex $(0, -1)$, that passes through $(-2,3)$. Considering that quadratic functions in vertex form: $y = a(x-h)^2 + k$ where (h, k) is the vertex of the function, then, we have: $y = a(x-0)^2 - 1 \rightarrow y = ax^2 - 1$. In addition, put $(-2,3)$ in $y = ax^2 - 1$: $3 = a(-2)^2 - 1 \rightarrow 3 = 4a - 1 \rightarrow 4a = 4 \rightarrow a = 1$. Finally, $y = x^2 - 1$.

Method 2: Using the solution to the system of equations, $(-2,3)$ and $(1,0)$:

A. $(1,0)$: $\begin{cases} x - 2y = 4 \rightarrow 1 - 2(0) = 4 \rightarrow 1 = 4 \\ y - 3 = x \rightarrow 0 - 3 = 1 \rightarrow -3 = 1 \end{cases}$. It is enough that one of the points in the equation system does not satisfy. Then, this is not true.

B. $(1,0)$: $\begin{cases} y^3 - 1 = x \rightarrow (0)^3 - 1 = 1 \rightarrow -1 = 1 \\ 2y - x = 1 \rightarrow 2(0) - 1 = 1 \rightarrow -1 = 1 \end{cases}$, is not true.

C. $(1,0)$: $\begin{cases} x^2 - 1 = y \rightarrow (1)^2 - 1 = 0 \\ x + y = 1 \rightarrow 1 - 0 = 1 \end{cases}$, is true. Now, put in the second solution of system.

$(-2,3)$: $\begin{cases} x^2 - 1 = y \rightarrow (-2)^2 - 1 = 3 \\ x + y = 1 \rightarrow -2 + 3 = 1 \end{cases}$.

Since this satisfies both points in the system of equations, this is answer.

D. $(1,0)$: $\begin{cases} x - y = -5 \rightarrow 1 - 0 = -5 \rightarrow 1 = -5 \\ y - 1 = x^2 \rightarrow 0 - 1 = (1)^2 \rightarrow -1 = 1 \end{cases}$. This one is also not true.

E. $(1, 0)$: $\begin{cases} x^2 = y + 1 \rightarrow 1^2 = 0 + 1 \\ x - y = 1 \rightarrow 1 - 0 = 1 \end{cases}$, is true. Now, put in the second solution of system.

$(-2, 3)$: $\begin{cases} x^2 = y + 1 \rightarrow (-2)^2 = 3 + 1 \\ x - y = 1 \rightarrow -2 - 3 = -5 \end{cases}$, it is incorrect!

15) Choice C is correct

Since the order doesn't matter, we need to use a combination formula where n is 7 and r is 4. Then:

$$\frac{n!}{r!(n-r)!} = \frac{7!}{4!(7-4)!} = \frac{7!}{4!(3)!} = \frac{7 \times 6 \times 5 \times 4!}{4!(3)!} = \frac{7 \times 6 \times 5}{3 \times 2 \times 1} = 35.$$

16) Choice E is correct

$$average = \frac{sum\ of\ terms}{number\ of\ terms} \rightarrow 20 = \frac{13+15+20+x}{4} \rightarrow 80 = 48 + x \rightarrow x = 32.$$

17) Choice C is correct

There can be 0, 1, or 2 solutions to a quadratic equation. In standard form, a quadratic equation is written as: $ax^2 + bx + c = 0$.

For the quadratic equation, the expression $b^2 - 4ac$ is called the discriminant. If the discriminant is positive, there are 2 distinct solutions for the quadratic equation. If the discriminant is 0, there is one solution for the quadratic equation and if it is negative the equation does not have any solutions.

To find the number of solutions for $x^2 = 4x - 3$, first, rewrite it as: $x^2 - 4x + 3 = 0$.

Find the value of the discriminant: $b^2 - 4ac = (-4)^2 - 4(1)(3) = 16 - 12 = 4$.

Since the discriminant is positive, the quadratic equation has two distinct solutions.

18) Choice E is correct

Since $f(x)$ is a linear function with a negative slope, then when $x = -2$, $f(x)$ is maximum and when $x = 3$, $f(x)$ is minimum. Then the ratio of the minimum value to the maximum value of the function is: $\frac{f(3)}{f(-2)} = \frac{-3(3)+1}{-3(-2)+1} = \frac{-8}{7} = -\frac{8}{7}$.

19) Choice C is correct

Let $x =$ the number of comic books and $y =$ the number of notebooks. From the question, we know that:

$$x = 4y + 5$$
$$x + y = 45$$

Substitute $4y + 5$ for the x in the second equation: $(4y + 5) + y = 45$. Combine like terms: $5y + 5 = 45$. Solve for y: $5y + 5 = 45 \rightarrow 5y = 40 \rightarrow y = 8$. Now, substitute 8 in for the y value: $x + 8 = 45$. Solve for x: $x = 37$.

20) Choice E is correct

If the value of $|x - 3| + 3$ is equal to 0, then $|x - 3| + 3 = 0$. Subtracting 3 from both sides of this equation gives $|x - 3| = -3$. The expression $|x - 3|$ on the left side of the equation is the absolute value of $x - 3$, and the absolute value can never be a negative number. Thus $|x - 3| = -3$ has no solution. Therefore, there are no values for x for which the value of $|x - 3| + 3$ is equal to 0.

21) Choice A is correct

Let x be the number of years. Therefore, \$2,000 per year equals $2{,}000x$. Starting from \$26,000 annual salary means you should add that amount to $2{,}000x$. Income more than that is: $I > 2{,}000x + 26{,}000$.

22) Choice C is correct

Multiply numerators and denominators: $\frac{3x+6}{x+5} \times \frac{x+5}{x+2} = \frac{(3x+6)(x+5)}{(x+5)(x+2)}$.

Cancel the common factor: $\frac{(3x+6)(x+5)}{(x+5)(x+2)} = \frac{(3x+6)}{(x+2)}$.

Factor $3x + 6 = 3(x + 2)$. Then: $\frac{3(x+2)}{(x+2)} = 3$.

23) Choice D is correct

In the figure angle A is labeled $(3x - 2)$ and it measures 37.

Thus, $3x - 2 = 37$ and $3x = 39$ or $x = 13$. That means that angle B, which is labeled $(5x)$, must measure $5 \times 13 = 65$. Since the three angles of a triangle must add up to 180,

$37 + 65 + y - 8 = 180$, then: $y + 94 = 180 \rightarrow y = 180 - 94 = 86$.

24) Choice C is correct

From the exponent rule: $(xy)^a = x^a \times y^a$. We have: $(3\alpha^4)^3 = 3^3 \times (\alpha^4)^3$. From this leads to: $(x^a)^b = x^{a \times b}$. Then: $(\alpha^4)^3 = \alpha^{4\times3} = \alpha^{12}$. Therefore:

$$(3\alpha^4)^3 = 3^3 \times (\alpha^4)^3 = 27\alpha^{12}.$$

25) Choice E is correct

The two radical parts are not the same. First, we need to simplify the $2\sqrt{32}$. Then: $2\sqrt{32} = 2\sqrt{16 \times 2} = 2(\sqrt{16})(\sqrt{2}) = 8\sqrt{2}$. Now, combine like terms: $2\sqrt{32} + 2\sqrt{2} = 8\sqrt{2} + 2\sqrt{2} = 10\sqrt{2}$.

26) Choice C is correct

Multiply by the conjugate: $\frac{\sqrt{12}+3}{\sqrt{12}+3} \rightarrow \frac{6}{\sqrt{12}-3} \times \frac{\sqrt{12}+3}{\sqrt{12}+3}$.

$$(\sqrt{12} - 3)(\sqrt{12} + 3) = 3, \text{ then: } \frac{6}{\sqrt{12}-3} \times \frac{\sqrt{12}+3}{\sqrt{12}+3} = \frac{6(\sqrt{12}+3)}{3} = 2(\sqrt{12} + 3).$$

27) Choice C is correct

Let x be the smallest number. Then, these are the numbers: $x, x+1, x+2, x+3, x+4$.

$average = \frac{sum\ of\ terms}{number\ of\ terms} \rightarrow 36 = \frac{x+(x+1)+(x+2)+(x+3)+(x+4)}{5} \rightarrow 36 = \frac{5x+10}{5} \rightarrow$

$180 = 5x + 10 \rightarrow 170 = 5x \rightarrow x = 34$.

28) Choice A is correct

$2x - 5 \geq 3x - 1$, add 5 to both sides: $2x - 5 + 5 \geq 3x - 1 + 5 \rightarrow 2x \geq 3x + 4$. Subtract $3x$ from both sides: $2x - 3x \geq 3x + 4 - 3x \rightarrow -x \geq +4$, multiply both side by -1 (Reverse the inequality): $(-x)(-1) \geq (+4)(-1) \rightarrow x \leq -4$. Only -2 is greater than -4.

29) Choice D is correct

Given the two equations, substitute the numerical value of a into the second equation to solve for x. $a = \sqrt{3}$, and $4a = \sqrt{4x}$.

Substituting the numerical value for a into the equation with x is as follows.

$4(\sqrt{3}) = \sqrt{4x}$, from here, distribute the 4. $4\sqrt{3} = \sqrt{4x}$.

Now square both sides of the equation. $(4\sqrt{3})^2 = (\sqrt{4x})^2$.

Remember to square both terms within the parentheses. Also, recall that squaring a square root sign cancels them out. $4^2(\sqrt{3})^2 = 4x$, $16(3) = 4x$, $48 = 4x$, $x = 12$.

30) Choice D is correct

First, we have: $3x(x+2) = 3x^2 + 6x$ and $(x+3)(x+1) = x^2 + 4x + 3$. Add two expressions:

$$(3x^2 + 6x) + (x^2 + 4x + 3) = 4x^2 + 10x + 3.$$

31) Choice C is correct

First square both sides of the equation to get $4m - 3 = m^2$.

Subtracting both sides by $4m - 3$ gives us the equation $m^2 - 4m + 3 = 0$.

Here you can solve the quadratic equation by factoring to get $(m-1)(m-3) = 0$.

For the expression $(m-1)(m-3)$ to equal zero, $m = 1$ or $m = 3$.

32) Choice C is correct

Let x be the number. Write the equation and solve for x: $(28 - x) \div x = 3$.

Multiply both sides by x. $(28 - x) = 3x$, then add x both sides. $28 = 4x$, now divide both sides by 4: $x = 7$.

33) Choice B is correct

The sum of supplement angles is 180. Let x be that angle. Therefore, $x + 9x = 180$.

$10x = 180$, divide both sides by 10: $x = 18$.

34) Choice A is correct

Find the factors of the denominators: $\frac{3x}{x^2-36} = \frac{3x}{(x-6)(x+6)}$ and $\frac{4}{2x-12} = \frac{4}{2(x-6)}$.

Since the factor $(x - 6)$ is common in both denominators, the least common denominator is: $2(x - 6)(x + 6) = 2x^2 - 72$.

35) Choice D is correct

$h(x) = \sqrt{x} + 3 \rightarrow y = \sqrt{x} + 3$, replace all x's with y and all y's with x.

$$x = \sqrt{y} + 3 \rightarrow x - 3 = \sqrt{y} \rightarrow (x - 3)^2 = \left(\sqrt{y}\right)^2 \rightarrow x^2 - 6x + 9 = y \rightarrow$$

$h^{-1}(x) = x^2 - 6x + 9$

36) Choice D is correct

Substituting 6 for x and 14 for y in $y = nx + 2$ gives $14 = (n)(6) + 2$,

which gives $n = 2$. Hence, $y = 2x + 2$. Therefore, when $x = 10$, the value of y is:

$$y = (2)(10) + 2 = 22.$$

37) Choice B is correct

6% of the volume of the solution is alcohol. Let x be the volume of the solution.

Then: 6% of $x = 24ml \rightarrow 0.06x = 24 \rightarrow x = 24 \div 0.06 = 400$.

38) Choice B is correct

$average = \frac{sum\ of\ terms}{number\ of\ terms}$. The sum of the weight of all girls is: $18 \times 56 = 1{,}008kg$.

The sum of the weight of all boys is: $32 \times 62 = 1{,}984kg$. The sum of the weight of all students is: $1{,}008 + 1{,}984 = 2{,}992kg$, $average = \frac{2{,}992}{50} = 59.84$.

39) Choice D is correct

We know that a scatter plot shows a positive trend if y tends to increase as x increases. By comparing that to the scatter plot, the trend is positive, and the outlier is $(8,1)$.

40) Choice D is correct

$$0.6x = (0.3) \times 20 \rightarrow x = 10 \rightarrow (x+3)^2 = (13)^2 = 169$$

41) Choice D is correct

A zero of a function corresponds to an x −intercept of the graph of the function in the xy −plane. Therefore, the graph of the function $g(x)$, which has three distinct zeros, must have three x −intercepts. Only the graph in choice D has three x −intercepts.

42) Choice E is correct

$(2.9 \times 10^6) \times (2.6 \times 10^{-5}) = (2.9 \times 2.6) \times (10^6 \times 10^{-5}) = 7.54 \times \left(10^{6+(-5)}\right) = 7.54 \times 10^1$.

43) Choice A is correct

Let x be the integer. Then: $2x - 5 = 73$, add 5 both sides: $2x = 78$, divide both sides by 2: $x = 39$.

44) Choice C is correct

Let x be all expenses, then:

$$\frac{22}{100}x = 660 \rightarrow x = \frac{100 \times 660}{22} = 3{,}000,$$

Mr. Green spent on his rent: $\frac{27}{100} \times 3{,}000 = 810$.

45) Choice A is correct

If $f(x) = 3x + 4(x+1) + 2$, then find $f(4x)$ by substituting $4x$ for every x in the function. This gives: $f(4x) = 3(4x) + 4(4x+1) + 2$.

It simplifies to: $f(4x) = 3(4x) + 4(4x+1) + 2 = 12x + 16x + 4 + 2 = 28x + 6$.

46) Choice C is correct

To solve by elimination, we want to eliminate one of the variables by adding or subtracting the two equations.

To eliminate y, we can multiply the second equation by 2 to match the coefficient of y from the first equation:

$6x - 2y = 14$ ... Multiplying the 2nd equation by 2.

Now, if we add the first equation to the modified second equation, we can eliminate y:

$$2x + 6x + 2y - 2y = 2 + 14$$

Combining like terms:

$8x = 16$ Divide both sides by 8: $x = 2$

Now, plug $x = 2$ into one of the original equations to solve for y. Using the first equation:

$2(2) + 2y = 2$ $4 + 2y = 2$ Subtract 4 from both sides: $y = -2$ Divide both sides by 2: $y = -1$

So, the solution to the system of equations is $x = 2$ and $y = -1$, which corresponds to the ordered pair $(2, -1)$.

Therefore, the correct answer from the given options is: $C. (2, -1)$.

47) Choice C is correct

First, find the equation of the line. All lines through the origin are of the form $y = mx$, so the equation is $y = \frac{1}{3}x$. Of the given choices, only choice C (9,3), satisfies this equation:

$$y = \frac{1}{3}x \rightarrow 3 = \frac{1}{3}(9) = 3.$$

48) Choice E is correct

$(3n^2 + 2n + 6) - (2n^2 - 4)$. Add like terms together: $3n^2 - 2n^2 = n^2$, $2n$ doesn't have like terms. $6 - (-4) = 10$, combine these terms into one expression to find the answer:

$$n^2 + 2n + 10.$$

49) Choice C is correct

Simplify and solve for x in the equation.

$$4(x + 1) = 6(x - 4) + 20 \rightarrow 4x + 4 = 6x - 24 + 20,\ 4x + 4 = 6x - 4.$$

Subtract $4x$ from both sides: $4 = 2x - 4$, add 4 to both sides: $8 = 2x$, $x = 4$.

50) Choice D is correct

$$3x + 10 = 46 \rightarrow 3x = 46 - 10 \rightarrow 3x = 36 \rightarrow x = \frac{36}{3} = 12.$$

51) Choice A is correct

To rewrite $\frac{1}{\frac{1}{x-5}+\frac{1}{x+4}}$, first simplify $\frac{1}{x-5} + \frac{1}{x+4}$.

$$\frac{1}{x-5} + \frac{1}{x+4} = \frac{1(x+4)}{(x-5)(x+4)} + \frac{1(x-5)}{(x+4)(x-5)} = \frac{(x+4)+(x-5)}{(x+4)(x-5)}.$$

Then: $\frac{1}{\frac{1}{x-5}+\frac{1}{x+4}} = \frac{1}{\frac{(x+4)+(x-5)}{(x+4)(x-5)}} = \frac{(x-5)(x+4)}{(x-5)+(x+4)}$. (Remember, $\frac{1}{\frac{1}{x}} = x$)

This result is equivalent to the expression in choice A.

52) Choice B is correct

Since $(0,0)$ is a solution to the system of inequalities, substituting 0 for x and 0 for y in the given system must result in two true inequalities. After this substitution, $y < c - x$ becomes $0 < c$, and $y > x + b$ becomes $0 > b$. Hence, a is positive and b is negative. Therefore, $c > b$.

53) Choice D is correct

To solve for x, isolate the radical on one side of the equation.

Divide both sides by 4. Then: $4\sqrt{2x+6} = 24 \rightarrow \frac{4\sqrt{2x+6}}{4} = \frac{24}{4} \rightarrow \sqrt{2x+6} = 6$.

Square both sides: $\left(\sqrt{(2x+6)}\right)^2 = 6^2$. Then: $2x + 6 = 36 \rightarrow 2x = 30 \rightarrow x = 15$.

Substitute x by 15 in the original equation and check the answer:

$$x = 15 \rightarrow 4\sqrt{2(15)+6} = 4\sqrt{36} = 4(6) = 24.$$

Therefore, the value of 15 for x is correct.

54) Choice B is correct

The input value is 5. Then: $x = 5$.

$$f(x) = x^2 - 3x \rightarrow f(5) = 5^2 - 3(5) = 25 - 15 = 10.$$

55) Choice B is correct

To solve this problem, first recall the equation of a line: $y = mx + b$.

Where $m =$ slope. $y = y$ −intercept.

Remember that slope is the rate of change that occurs in a function and that the y −intercept is the y value corresponding to $x = 0$.

Since the height of John's plant is 5 inches tall when he gets it. Time (or x) is zero. The plant grows 3 inches per year. Therefore, the rate of change of the plant's height is 3. The y −intercept represents the starting height of the plant which is 5 inches.

56) Choice D is correct

Multiplying each side of $\frac{4}{x} = \frac{12}{x-8}$ by $x(x-8)$ gives $4(x-8) = 12(x)$, distributing the 4 over the values within the parentheses yields $x - 8 = 3x$. Then, $x = -4$.

Therefore, the value of $\frac{x}{2} = \frac{-4}{2} = -2$.

57) Choice A is correct

The amount of money for x bookshelf is: $200x$, then, the total cost of all bookshelves is equal to: $200x + 600$, the total cost, in dollars, per bookshelf is:

$$\frac{Total\ cost}{number\ of\ items} = \frac{200x + 600}{x}$$

58) Choice B is correct

It is given that $g(5) = 4$. Therefore, to find the value of $f(g(5))$, then:

$f(g(5)) = f(4) = 6$.

59) Choice C is correct

$6 \blacksquare 28 = \sqrt{6^2 + 28} = \sqrt{36 + 28} = \sqrt{64} = 8$.

60) Choice E is correct

Solve the system of equations by elimination method.

$\begin{array}{l} 3x - 4y = -20 \\ -x + 2y = 10 \end{array}$. Multiply the second equation by 3, then add it to the first equation.

$\begin{array}{l} 3x - 4y = -20 \\ 3(-x + 2y = 10) \end{array} \rightarrow \begin{array}{l} 3x - 4y = -20 \\ -3x + 6y = 30 \end{array} \rightarrow$ add the equations $2y = 10 \rightarrow y = 5$.

Algebra I Practice Tests 2 Explanations

1) Choice A is correct

$$\left(\frac{f}{g}\right)(x) = \frac{f(x)}{g(x)} = \frac{2x+2}{x^2+4x}$$

2) Choice D is correct

The equation of a line is: $y = mx + b$, where m is the slope and b is the y –intercept.

First find the slope: $m = \frac{y_2-y_1}{x_2-x_1} = \frac{15-(-5)}{8-3} = \frac{20}{5} = 4$. Then, we have: $y = 4x + b$.

Choose one point and plug in the values of x and y in the equation to solve for b.

Let's choose the point $(3, -5)$.

$$y = 4x + b \rightarrow -5 = 4(3) + b \rightarrow -5 = 12 + b \rightarrow b = -17.$$

The equation of the line is: $y = 4x - 17$.

3) Choice D is correct

Use FOIL (First, Out, In, Last) method:

$$(5x + 2y)(2x - y) = 10x^2 - 5xy + 4xy - 2y^2 = 10x^2 - xy - 2y^2$$

4) Choice D is correct

To solve absolute value equations, write two equations. $x - 10$ could be positive 4, or negative 4. Therefore, $x - 10 = 4 \rightarrow x = 14$, $x - 10 = -4 \rightarrow x = 6$.

Find the product of solutions: $6 \times 14 = 84$

5) Choice A is correct

The vertical asymptote of a rational function $y = f(x)$ as $f(x) = \frac{h(x)}{g(x)}$ is a vertical line $x = k$ when, $g(x) = 0$. Solve the equation: $x^2 + 2 = 0 \rightarrow x^2 = -2$. The denominator is not zero for any value. Therefore, this graph does not have a vertical asymptote.

6) Choice A is correct

The equation of a line in slope intercept form is: $y = mx + b$. Solve for y.

$$4x - 2y = 6 \rightarrow -2y = 6 - 4x \rightarrow y = (6 - 4x) \div (-2) \rightarrow y = 2x - 3.$$

The slope is 2. The slope of the line perpendicular to this line is:

$m_1 \times m_2 = -1 \rightarrow 2 \times m_2 = -1 \rightarrow m_2 = -\frac{1}{2}$.

7) Choice C is correct

Plug in the value of x and y. $x = 3$ and $y = -2$.

$6(x - 2y) + (2 - x)^2 = 6(3 - 2(-2)) + (2 - 3)^2 = 6(3 + 4) + (-1)^2 = 42 + 1 = 43$.

8) Choice D is correct

Write the ordered pairs from the scatter plot and compare them with the data in the tables as follows: (1,1000), (2,800), (3,600), (4,500), (5,400), (6,350), (7,300).

9) Choice C is correct

First find the value of b, and then find $f(3)$. Since $f(2) = 35$, substituting 2 for x and 35 for $f(x)$ gives $35 = b(2)^2 + 15 \rightarrow 35 = 4b + 15$.

Solving this equation gives $b = 5$. Thus $f(x) = 5x^2 + 15$,

$$f(3) = 5(3)^2 + 15 \rightarrow f(3) = 45 + 15 \rightarrow f(3) = 60$$

10) Choice B is correct

The vertical asymptote of a rational function $y = f(x)$ as $f(x) = \frac{1}{g(x)}$ is a vertical line. $x = k$ when, $g(x) = 0$. Solve the equation of $x^3 + 8 = 0$.

$$x^3 + 8 = 0 \rightarrow x^3 = -8 \rightarrow x = -2.$$

11) Choice D is correct

For domain: find non-negative values for radicals: $5x - 15 \geq 0$.

Domain of functions: $5x - 15 \geq 0 \rightarrow 5x \geq 15 \rightarrow x \geq 3$.

Domain of the function $y = 7\sqrt{5x - 15} + 5$: $x \geq 3$.

12) Choice D is correct

Solve for the sum of five numbers.

$average = \frac{sum\ of\ terms}{number\ of\ terms} \rightarrow 26 = \frac{sum\ of\ 5\ numbers}{5}$

$\rightarrow sum\ of\ 5\ numbers = 26 \times 5 = 130$.

The sum of 5 numbers is 130. If a sixth number 42 is added, then the sum of 6 numbers is:

$$130 + 42 = 172.$$

The new average is: $\frac{sum\ of\ 6\ numbers}{6} = \frac{172}{6} = 28.666 \approx 28.67$.

13) Choice D is correct

The rate of construction company $= \frac{30\ cm}{1\ min} = 30\ cm/min.$

Height of the wall after 40 minutes $= \frac{30\ cm}{1\ min} \times 40\ min = 1{,}200\ cm.$

Let x be the height of wall, then $\frac{3}{4}x = 1{,}200\ cm \rightarrow x = \frac{4 \times 1{,}200}{3} \rightarrow x = 1{,}600\ cm = 16\ m.$

14) Choice A is correct

$x - 2 \geq 3 \rightarrow x \geq 3 + 2 \rightarrow x \geq 5$ or $x - 2 \leq -3 \rightarrow x \leq -3 + 2 \rightarrow x \leq -1.$

Then, solution is: $x \leq -1 \cup x \geq 5.$

15) Choice C is correct

We know that the graph $y = f(x - k); k > 0$, is shifted k units to the right of the graph $y = f(x)$ and if $k < 0$, is shifted k units to the left. Now, to move $f(x) = -x^2$, to the right by two units, we have: $f(x - 2) = -(x - 2)^2$.
Therefore: $g(x) = -(x - 2)^2$.

16) Choice C is correct

When 5 times the number x is added to 10, the result is $10 + 5x$. Since this result is equal to 35, the equation $10 + 5x = 35$ is true. Subtracting 10 from each side of $10 + 5x = 35$ gives $5x = 25$, and then dividing both sides by 5 gives $x = 5$. Therefore, 3 times x added to 6, or $6 + 3x$, is equal to $6 + 3(5) = 21$.

17) Choice B is correct

Finding g in term of h, simply means "solve the equation for g". To solve for g, isolate it on one side of the equation. Since g is on the left-hand side, just keep it there. Subtract both sides by $3h$. $3h + g - 3h = 8h + 4 - 3h$.

And simplifying makes the equation $g = 5h + 4$.

18) Choice B is correct

Isolate and solve for x. $\frac{2}{3}x + \frac{1}{6} = \frac{1}{2} \rightarrow \frac{2}{3}x = \frac{1}{2} - \frac{1}{6} \rightarrow \frac{2}{3}x = \frac{1}{3}$. Multiply both sides by the reciprocal of the coefficient of x. $(\frac{3}{2})\frac{2}{3}x = \frac{1}{3}(\frac{3}{2}) \rightarrow x = \frac{3}{6} = \frac{1}{2}$.

19) Choice B is correct

Use simple interest formula: $I = prt$ (I = interest, p = principal, r = rate, t = time).

$I = (12{,}000)(0.045)(2) = 1{,}080.$

20) Choice D is correct

Use the combination formula: $\frac{n!}{r!(n-r)!} \rightarrow \frac{12!}{4!(12-4)!} = \frac{12!}{4!8!} = 495.$

21) Choice D is correct

Simplify. $7x^2y^3(2x^2y)^3 = 7x^2y^3(8x^6y^3) = 56x^8y^6$

22) Choice E is correct

Frist factor the function: $f(x) = x^3 + 7x^2 + 12x = x(x+3)(x+4)$.

To find the zeros, $f(x)$ should be zero: $f(x) = x(x+3)(x+4) = 0$.

Therefore, the zeros are: $x = 0$, $(x+3) = 0 \rightarrow x = -3$, $(x+4) = 0 \rightarrow x = -4$.

23) Choice E is correct

Consider the variable part to be $a(a+b)^2$, then this expression has two like terms with coefficients -15 and 23. Then: $-15a(a+b)^2 + 23a(a+b)^2 = 8a(a+b)^2$.

24) Choice E is correct

Solve for x. $\sqrt{3x} = \sqrt{y}$. Square both sides of the equation:

$$(\sqrt{3x})^2 = (\sqrt{y})^2 \rightarrow 3x = y \rightarrow x = \frac{y}{3}.$$

25) Choice C is correct

To solve the inequality, bring the variable x to one side by adding or subtracting.

Then: $8x - 4 \geq -2x + 16 \rightarrow 8x + 2x - 4 \geq -2x + 16 + 2x \rightarrow 10x - 4 + 4 \geq 16 + 4$

$\rightarrow 10x \geq 20 \rightarrow x \geq \frac{20}{10} \rightarrow x \geq 2$, as the variable is greater than or equal to 2.

26) Choice E is correct

$g(x) = -4$, then:

$$f(g(x)) \rightarrow f(-4) = 2(-4)^3 + 5(-4)^2 + 2(-4) = -128 + 80 - 8 = -56.$$

27) Choice B is correct

Plug in 104 for F and then solve for C.

$$C = \frac{5}{9}(F - 32) \rightarrow C = \frac{5}{9}(104 - 32) \rightarrow C = \frac{5}{9}(72) = 40$$

28) Choice D is correct

$average = \frac{sum\ of\ terms}{number\ of\ terms} \rightarrow$ (Average of 6 numbers) $14 = \frac{sum\ of\ numbers}{6} \rightarrow$

sum of 6 numbers is $14 \times 6 = 84$, (average of 4 numbers) $10 = \frac{sum\ of\ numbers}{4} \rightarrow$

sum of 4 numbers is $10 \times 4 = 40$.

$sum\ of\ 6\ numbers - sum\ of\ 4\ numbers = sum\ of\ 2\ numbers,\ 84 - 40 = 44.$

$$average\ of\ 2\ numbers = \frac{44}{2} = 22$$

29) Choice A is correct

First, find the number. Let x be the number. Write the equation and solve for x. 150% of a number is 75, then:

$$1.5 \times x = 75 \rightarrow x = 75 \div 1.5 \rightarrow x = 50,\ 80\% \text{ of } 50 \text{ is: } 0.8 \times 50 = 40.$$

30) Choice A is correct

The function is not defined for $x = -2$, as this value would result is division by zero.

Hence the domain of $f(x)$ is "all real numbers except -2.

Range: no matter how large or small x becomes, $f(x)$ will never be equal to zero.

So, the range of $f(x)$ is "all real numbers except zero".

31) Choice E is correct

Solve for y. $4x - 2y = 12 \rightarrow -2y = 12 - 4x \rightarrow y = 2x - 6$. The slope of the line is 2.

32) Choice B is correct

The population has increased by 10% and 20%. A 10% increase changes the population to 110% of the original population. For the second increase, multiply the result by 120%.

$(1.10) \times (1.20) = 1.32 \rightarrow 132\%$. 32 percent of the population has increased after two years.

33) Choice D is correct

If 20% of a number is 4, what is the number:

$$20\% \text{ of } x = 4 \rightarrow 0.2x = 4 \rightarrow x = 4 \div 0.2 \rightarrow x = 20.$$

34) Choice C is correct

Substituting x for y in the first equation. $5x + 2y = 3,\ 5x + 2(x) = 3,\ 7x = 3$.

Divide both side of $7x = 3$ by 3 gives $x = \frac{3}{7}$.

35) Choice D is correct

Apply exponent rules: $x^a \times x^b = x^{a+b}$.

Then: $3x^4u^5 . 5u^2 . 2x = 30x^5u^7$.

36) Choice B is correct

There are 5 floors, x rooms on each floor, and y chairs per room. If you multiply 5 floors by x, there are $5x$ rooms in the hotel. To get the number of chairs in the hotel, multiply $5x$ by y. $5xy$ is the number of chairs in the hotel.

37) Choice D is correct

If $\beta = 3\gamma$, then multiplying both sides by 12 gives $12\beta = 36\gamma$.

$\alpha = 2\beta$, thus $\alpha = 6\gamma$. Multiply both sides of the equation by 6 gives $6\alpha = 36\gamma$.

38) Choice E is correct

$$f(g(x)) = 2 \times \left(\frac{1}{x}\right)^3 + 2 = \frac{2}{x^3} + 2$$

39) Choice D is correct

The x −axis is the horizontal axis. The line intersects the x −axis at $(-3,0)$.

40) Choice B is correct

Find the common denominator and simplify the expression.

$$\sqrt{\frac{x^2}{2} + \frac{x^2}{16}} = \sqrt{\frac{8x^2}{16} + \frac{x^2}{16}} = \sqrt{\frac{9x^2}{16}} = \sqrt{\frac{9}{16}x^2} = \sqrt{\frac{9}{16}} \times \sqrt{x^2} = \frac{3}{4} \times x = \frac{3}{4}x.$$

41) Choice C is correct

Write the ratio of $5a$ to $2b$. $\frac{5a}{2b} = \frac{1}{10}$. Use cross multiplication and then simplify.

$$5a \times 10 = 2b \times 1 \rightarrow 50a = 2b \rightarrow a = \frac{2b}{50} = \frac{b}{25}.$$

Now, find the ratio of a to b. $\frac{a}{b} = \frac{\frac{b}{25}}{b} \rightarrow \frac{b}{25} \div b = \frac{b}{25} \times \frac{1}{b} = \frac{b}{25b} = \frac{1}{25}$.

42) Choice A is correct

Plug in the value of x in the equation and solve for y: $2y = \frac{2x^2}{3} + 6 \rightarrow 2y = \frac{2(9)^2}{3} + 6 \rightarrow$ $2y = \frac{2(81)}{3} + 6 \rightarrow 2y = 54 + 6 \rightarrow 2y = 60 \rightarrow y = 30$.

43) Choice D is correct

Since $N = 6$, substitute 6 for N in the equation $\frac{x-3}{5} = N$, which gives $\frac{x-3}{5} = 6$. Multiplying both sides of $\frac{x-3}{5} = 6$ by 5 gives $x - 3 = 30$ and then adding 3 to both sides of $x - 3 = 30$. Then, $x = 33$.

44) Choice C is correct

$b^{\frac{m}{n}} = \sqrt[n]{b^m}$ For any positive integers m and n. Thus, $b^{\frac{3}{5}} = \sqrt[5]{b^3}$.

45) Choice C is correct

The total number of pages read by Sara is 3 (Hours she spent reading) multiplied by her rate of reading: $\frac{Npages}{hour} \times 3hours = 3N$.

Similarly, the total number of pages read by Mary is 4 (Hours she spent reading) multiplied by her rate of reading: $\frac{Mpages}{hour} \times 4hours = 4M$ the total number of pages read by Sara and Mary is the sum of the total number of pages read by Sara and the total number of pages read by Mary: $3N + 4M$.

46) Choice C is correct

By comparing ordered pairs of scatter plot and tables, you can see that the third choice is best: $(1,37)$, $(2,38)$, $(3,27)$, $(4,39)$, $(5,38)$, $(6,39)$, $(7,40)$, $(8,39)$, $(9,41)$, $(10,41)$.

47) Choice B is correct

There are several ways to graph a line. For this one, it is convenient to find the x intercept and the y intercept. To find the x intercept, substitute 0 for y, resulting is the equation $4x = 8$. Dividing both sides by 4, gives $x = 2$. Plot the point $(2,0)$. To find the y intercept, substitute 0 for x, resulting is the equation $2y = 8$. Dividing both sides by 2, gives $y = 4$. Plot the point $(0,4)$. Drawing the line through the two points gives the graph in B.

48) Choice B is correct

Solving Systems of Equations by Elimination: multiply the first equation by (-2), then add it to the second equation.

$$\begin{array}{l} -2(2x + 5y = 11) \\ \underline{4x - 2y = -14} \end{array} \rightarrow \begin{array}{l} -4x - 10y = -22 \\ 4x - 2y = -14 \end{array} \rightarrow -12y = -36 \rightarrow y = 3.$$

Plug in the value of y into one of the equations and solve for x.

$$2x + 5(3) = 11 \rightarrow 2x + 15 = 11 \rightarrow 2x = -4 \rightarrow x = -2.$$

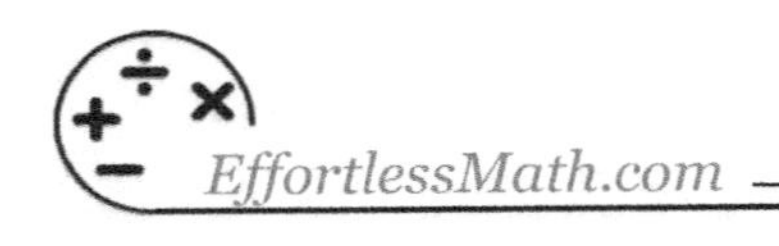

49) Choice C is correct

A. $|4-2| = |2| = 2$

B. $|2-4| = |-2| = 2$

C. $|-2-4| = |-6| = 6$

D. $|2-4| - |4-2| = |-2| - |2| = 2 - 2 = 0$

E. $|2-4| + |4-2| = |-2| + |2| = 2 + 2 = 4$

Choice C is the largest number.

50) Choice B is correct

Based on the triangle similarity theorem: $\frac{a}{a+b} = \frac{c}{3} \rightarrow c = \frac{3a}{a+b} = \frac{3\sqrt{3}}{3\sqrt{3}} = 1 \rightarrow$ The area of shaded region is: $\left(\frac{c+3}{2}\right)(b) = \frac{4}{2} \times 2\sqrt{3} = 4\sqrt{3}$.

51) Choice D is correct

This can be ascertained by putting the ordered pairs of scatter plot in each equation.
For example, put $(1, -2)$ in the equations:
A. $(1,-2) \rightarrow y = x \rightarrow -2 = 1$, is not true.
B. $(1,-2) \rightarrow y = -x \rightarrow -2 = -(1) \rightarrow -2 = -1$, this is not true.
C. $(1,-2) \rightarrow y = x + 1 \rightarrow -2 = 1 + 1 \rightarrow -2 = 2$, this one is also not true.
D. $(1,-2) \rightarrow y = -2x \rightarrow -2 = -2(1) \rightarrow -2 = -2$, this is true.
E. $(1,-2) \rightarrow y = 2x \rightarrow -2 = 2(1) \rightarrow -2 = 2$, this is not true.
Similarly, put the other points of the scatter plot on this equation.

52) Choice D is correct

If $x - a$ is a factor of $g(x)$, then $g(a)$ must equal 0. Based on the table $g(2) = 0$. Therefore, $x - 2$ must be a factor of $g(x)$.

53) Choice C is correct

To solve this problem first solve the equation for c: $\frac{c}{b} = 2$.

Multiply by b on both sides. Then: $b \times \frac{c}{b} = 2 \times b \rightarrow c = 2b$. Now to calculate $\frac{4b}{c}$, substitute the value for c into the denominator and simplify.

$$\frac{4b}{c} = \frac{4b}{2b} = \frac{4}{2} = \frac{2}{1} = 2$$

54) Choice B is correct

$x + 5 = 8 \rightarrow x = 8 - 5 \rightarrow x = 3$, $2y - 1 = 5 \rightarrow 2y = 6 \rightarrow y = 3$,

$xy + 15 = 3 \times 3 + 15 = 24$.

55) Choice D is correct

When you plug the points in for the x and y, only choice D $(1,0)$ satisfies both inequalities. For this purpose, we have:

A. $(-1,0)$: $\begin{cases} y < \frac{1}{5}x + 3 \rightarrow 0 < \frac{1}{5}(-1) + 3 = \frac{14}{5} \\ 2y + x \geq 1 \rightarrow 2 \times 0 + (-1) = -1 \geq 1 \end{cases}$, this is not true.

B. $(0,4)$: $\begin{cases} y < \frac{1}{5}x + 3 \rightarrow 4 < \frac{1}{5}(0) + 3 = 3 \\ 2y + x \geq 1 \rightarrow 2(4) + 0 = 8 \geq 1 \end{cases}$, this is not true.

C. $(1,4)$: $\begin{cases} y < \frac{1}{5}x + 3 \rightarrow 4 < \frac{1}{5}(1) + 3 = \frac{16}{5} \\ 2y + x \geq 1 \rightarrow 2(4) + 1 = 9 \geq 1 \end{cases}$, this is not true.

D. $(1,0)$: $\begin{cases} y < \frac{1}{5}x + 3 \rightarrow 0 < \frac{1}{5}(1) + 3 = \frac{16}{5} \\ 2y + x \geq 1 \rightarrow 2 \times 0 + 1 = 1 \geq 1 \end{cases}$, this is true.

E. $(0,-1)$: $\begin{cases} y < \frac{1}{5}x + 3 \rightarrow -1 < \frac{1}{5}(0) + 3 = 3 \\ 2y + x \geq 1 \rightarrow 2 \times (-1) + 0 = -2 \geq 1 \end{cases}$, this is not true.

56) Choice E is correct

The equation $\frac{a-b}{b} = \frac{10}{13}$ can be rewritten as $\frac{a}{b} - \frac{b}{b} = \frac{10}{13}$, from which it follows that:

$\frac{a}{b} - 1 = \frac{10}{13}$, or $\frac{a}{b} = \frac{10}{13} + 1 = \frac{23}{13}$.

57) Choice C is correct

First write the equation in slope-intercept form. Add $2x$ to both sides to get $6y = 2x + 24$. Now divide both sides by 6 to get $y = \frac{1}{3}x + 4$. The slope of this line is $\frac{1}{3}$, so any line that also has a slope of $\frac{1}{3}$ would be parallel to it. Only choice C has a slope of $\frac{1}{3}$.

58) Choice D is correct

Total score of first 12 students $= 12 \times 75 = 900$.

Total score of remaining 10 students $= 10 \times 85 = 850$.

Mean score of the whole class $= \frac{900+850}{22} \approx 79.5$.

59) Choice E is correct

Since a box of pens costs \$3, then $3p$ represents the cost of p boxes of pens. Multiplying this number times 1.085 will increase the cost by 8.5% for tax. Then add the \$6 shipping fee for the total: $1.085(3p) + 6$.

60) Choice D is correct

The rate of change (growth or x) is 8 per week: $40 \div 5 = 8$.

Since the plant grows at a linear rate, then the relationship between the height (y) of the plant and the number of weeks of growth (x) can be written as: $y(x) = 8x$.

Effortless Math's Algebra I Online Center

Effortless Math Online Algebra I Center offers a complete study program, including the following:

- ✓ Step-by-step instructions on how to prepare for the Algebra I test
- ✓ Numerous Algebra I worksheets to help you measure your math skills
- ✓ Complete list of Algebra I formulas
- ✓ Video lessons for all Algebra I topics
- ✓ Full-length Algebra I practice tests
- ✓ And much more…

No Registration Required.

Visit **EffortlessMath.com/Algebra1** to find your online Algebra I resources.

Receive the PDF version of this book or get another FREE book!

Thank you for using our Book!

Do you LOVE this book?

Then, you can get the PDF version of this book or another book absolutely FREE!

Please email us at:

info@EffortlessMath.com

for details.

Author's Final Note

I hope you enjoyed reading this book. You've made it through the book! Great job!

I would like to express my sincere appreciation for choosing this study guide to aid in your preparation for your Algebra I course. With a plethora of options available, I am grateful that you selected this book.

It took me years to write this study guide for the Algebra I because I wanted to prepare a comprehensive Algebra I study guide to help students make the most effective use of their valuable time while preparing for the final exam.

Over the course of my decade-long career teaching and tutoring math, I have compiled my personal notes and experiences into the creation of this study guide. It is my fervent hope that the information and lessons contained within these pages will assist you in achieving success on your Algebra I exam.

If you have any questions, please contact me at reza@effortlessmath.com and I will be glad to assist. Your feedback will help me to greatly improve the quality of my books in the future and make this book even better. Furthermore, I expect that I have made a few minor errors somewhere in this study guide. If you think this to be the case, please let me know so I can fix the issue as soon as possible.

If you enjoyed this book and found some benefit in reading this, I'd like to hear from you and hope that you could take a quick minute to post a review on the book's Amazon page. To leave your valuable feedback, please visit: amzn.to/3wNlPIV

Or scan this QR code.

I personally go over every single review, to make sure my books really are reaching out and helping students and test takers. Please help me help Algebra students, by leaving a review!

I wish you all the best in your future success!

Reza Nazari

Math teacher and author

Made in the USA
Monee, IL
01 October 2023

43708522R00155